REMNANTS OF ANCIENT LIFE

REMNANTS OF ANCIENT LIFE

THE NEW SCIENCE OF OLD FOSSILS

DALE E. GREENWALT

PRINCETON UNIVERSITY PRESS

PRINCETON & OXFORD

Published by Princeton University Press
41 William Street, Princeton, New Jersey 08540
99 Banbury Road, Oxford OX2 6JX

press.princeton.edu

Library of Congress Cataloging-in-Publication Data

Names: Greenwalt, Dale E., 1949– author.
Title: Remnants of ancient life : the new science of old fossils / Dale E. Greenwalt.
Description: Princeton, New Jersey : Princeton University Press, [2022] | Includes bibliographical references and index.
Identifiers: LCCN 2022013105 (print) | LCCN 2022013106 (ebook) | ISBN 9780691221144 (hardback) | ISBN 9780691221151 (ebook)
Subjects: LCSH: Biomolecules, Fossil. | DNA, Fossil—Analysis. | BISAC: NATURE / Fossils | SCIENCE / Earth Sciences / General
Classification: LCC QP517.F66 G74 2022 (print) | LCC QP517.F66 (ebook) | DDC 572/.786—dc23/eng/20220701
LC record available at https://lccn.loc.gov/2022013105
LC ebook record available at https://lccn.loc.gov/2022013106

British Library Cataloging-in-Publication Data is available

Editorial: Alison Kalett and Hallie Schaeffer
Production Editorial: Kathleen Cioffi
Text Design: Heather Hansen
Jacket Design: Chris Ferrante
Production: Jacqueline Poirier
Publicity: Sara Henning-Stout and Kate Farquhar-Thomson

Jacket art: *Encrinus lilliformis* illustrated in James Parkinson, *Organic Remains of a Former World*, vol. 2 (1808). Although there is no evidence that this particular specimen contains ancient pigments, the purple color was added here to connote the fact that similar fossils do contain ancient pigments. Courtesy of Oxford University Museum of Natural History.

This book has been composed in Arno Pro

Printed on acid-free paper. ∞

Printed in the United States of America

10 9 8 7 6 5 4 3 2 1

CONTENTS

ACKNOWLEDGMENTS

Remnants of Ancient Life had its origin in the discovery of a 46-million-year-old blood-engorged mosquito collected by the Constenius family of Whitefish, Montana. Their contribution is a wonderful example of the potential for amateur fossil enthusiasts to contribute to science. As a direct result of the publication of a description of that amazing fossil, I received an invitation from Kenneth De Baets (Friedrich-Alexander-University Erlangen-Nuremberg) and John Huntley (University of Missouri) to write a chapter on the fossil record of blood for a book entitled *The Evolution and Fossil Record of Parasitism*. The literature searches that I did for that article introduced me to the exciting and growing field of ancient biomolecules. But there are numerous other individuals whose contributions were critical to the publication of *Remnants of Ancient Life*.

I would like to thank the staff of the Paleobiology Department of the Smithsonian's National Museum of Natural History (NMNH), with special thanks to Conrad Labandeira, Curator of Fossil Arthropods, for sponsoring my position in the department as a Resident Research Associate. The nourishing environment at the museum makes it easy to envision new projects—such as a book about natural science that seeks to bring science to the nonscientists of the world. I would like to thank Richard Barclay, Selina Cole, Vera Korasidis, Tim Rose, Yulia Goreva, Sandra Siljeström, Finnegan Marsh, and Jon Wingerath, all from the NMNH, for conversations about ancient biomolecules and the fossils that contain them and/or their assistance in my own research with ancient biomolecules. Discussions with Bill Rember (University of Idaho), Stephen Godfrey and John Nance (Calvert Marine Museum), Chris Mason (AVROBio, Inc.), Derek Briggs (Yale University) and Klaus Wolkenstein (University of Göttingen) were also invaluable. I thank Alan Munro, Jill Warren, Rich Barclay (NMNH), Dong Ren (Capital Normal University, Beijing) and the Denver Museum of Nature and Science for photographs that appear in *Remnants of Ancient Life* as well as Yoho National Park, Glacier National Park and the NMNH for permission to photograph specimens.

My agent Laura Wood deserves particular credit and thanks for convincing me that most all of my initial assumptions about writing science for the lay public were wrong. It was a difficult lesson to learn but, ultimately, an invaluable one. Thanks are due to Derek Briggs (Yale University), Evan Saitta (University of Chicago) and Ginny Gray Johnson for early reviews of the manuscript and to Jill Warren and Ginny Gray Johnson for proofreading.

Thanks to Amanda Moon and James Brandt for edits of the book that greatly increased the clarity and comprehensibility of the text, and Julie Shawvan, who produced the index. The people at Princeton University Press were, of course, at many levels, critical to the publication of *Remnants of Ancient Life*. I want to thank Alison Kalett, who first recognized a glint of potential in the original manuscript, and Hallie Schaeffer (Assistant Editor, Biology) and Kathleen Cioffi (Senior Production Editor) and their teams, who did a very professional job of converting a manuscript into a book. Thanks also to Sara Henning Stout and Kate Farquhar-Thomson, the publicists for the book.

REMNANTS OF ANCIENT LIFE

A Chart of Geological Time Periods

Period	Start (millions of years ago)	End (millions of years ago)	In situ biomolecules
Quaternary	2.58	present	DNA (Tutankhamun's malaria)
Holocene	0.117	present	Coagulation factors (Ötzi)
Pleistocene	2.58	0.117	DNA (Yukon horse)
Cenozoic	66	present	
Miocene	23.03	5.33	Chlorophyll (leaf)
Oligocene	33.9	23.03	Chitin (Beetle)
Eocene	56	33.9	Heme (blood-engorged mosquito)
Cretaceous	145	66	Melanin pigment (*T. rex* feather)
Jurassic	201.3	145	Red pigment (Red-banded alga), Melanin (squid ink)
Triassic	251.9	201.3	
Permian	298.9	251.9	
Carboniferous	358.9	298.9	
Pennsylvanian	323.2	298.9	
Mississippian	358.9	323.2	Purple pigment (Crinoid)
Devonian	419.2	358.9	Cholesterol-related biomolecules (Crustacean)
Silurian	443.8	419.2	
Ordovician	485.4	443.8	
Cambrian	541	485.4	Chitin (*Vauxia*)
Ediacaran	635	541	Cholesterol-like molecule (*Dickensonia*)
Proterozoic	2500	541	Organic-walled microfossils

INTRODUCTION

Every summer, after 11 months of work in a laboratory buried deep in the bowels of the Smithsonian's National Museum of Natural History in Washington, DC, I have the chance to do what every paleobiologist dreams of doing when they decide to become a paleobiologist: fieldwork. The opportunity to find fossils of organisms that lived tens or even hundreds of millions of years ago casts a spell that is both potent and universal, as such fossils provide a road map to the origins and evolution of life on our planet. When we stare at a fossil, whether it is a billion-year-old stromatolite cemented into a Montana cliff or a *Tyrannosaurus rex* toe bone from the hills of Wyoming, we pause and reflect. Perhaps we think of our own very fleeting lifespan—not of our individual life but that of our species. Or perhaps we are drawn to the near infinite number of events that were required for the evolutionary processes that gave rise to the fossil. We may wonder how the organism died and how the fossil, whether made of stone, encased in amber, or mummified by desiccation, could have possibly survived these millions of years.

It was a specimen from the Kishenehn Formation in northwestern Montana, the fossil of a mosquito, that led me to ask a very different question. This was no ordinary mosquito. It was the beautifully preserved fossil of a blood-engorged mosquito—the first ever found. We have all watched as a mosquito pushes its proboscis though our skin, searches for a tiny blood vessel, and begins to transfer blood into its abdomen. If we are patient, we see the abdomen of the insect expand and darken. If we are quick, we see with what little force the blood-engorged mosquito can be smashed into unrecognizable fragments, a smear of blood spread over our skin. The chances that a blood-engorged mosquito, blown up like a taut balloon, would survive, intact, through the long and complex fossilization process are next to nothing. Examining this impossible specimen through my hand lens, I thought of Michael Crichton's *Jurassic Park*. Could there be DNA present? Perhaps even dinosaur DNA? No, of course not. The rocks were too young. But might some trace of blood, some ancient biomolecule that was once an integral part of the insect, have been preserved? Answering this question would lead to two unexpected events. First, when my colleagues and I published a paper that described the preservation of 46-million-year-old remnants of hemoglobin in the abdomen of the fossil, and I was interviewed by National Public Radio, the fossil became fleetingly famous—at least as famous as a fossil insect can be in our dinosaur-centric world. Second, I became engrossed in the rapidly growing science of ancient biomolecules—the study of DNA, protein, pigments, and other organic material that has been preserved across millions of years. This fascination led me to write this book, so I could share some of the field's awe-inspiring discoveries and explain how this focus on ancient biomolecules is completely changing the game of paleobiology.

For hundreds of years, paleobiologists have relied on a single tool with which to study, classify, and understand fossil organisms: comparative anatomy. It is a powerful and surprisingly discriminating instrument. The molars of modern humans and Neandertals can be easily distinguished. Muscles, tendons, and ligaments leave behind "scars" where they attached to bones, which can be used, for example, to establish ages of individuals of the same species. Through a histological examination of dinosaur bones, scientists have even been able to determine that a dinosaur was not only female but pregnant. In the past, phylogeny—through which we seek to understand the evolutionary relationships of one organism to another—has been based solely on the morphology of the fossilized remnants of extinct animals.

Now, though, we can peer into the past by examining several different kinds of ancient biomolecules. We have ancient DNA. And not just degraded fragments of DNA but entire ancient genomes: the very source of evolution. Access to ancient genes has already allowed us to study evolution at the molecular level; the oldest sample to date, nearly 1.8 million years old, has been used to trace the early Pleistocene evolution of rhinoceroses. Ancient DNA has also allowed scientists to synthesize ancient proteins and show that their function differed from their modern counterparts.

We also have ancient proteins, which are even older than ancient DNA. While most scientists agree that the oldest ancient protein sequences to date are about 3.8 million years old, there are data that suggest that sequenceable proteins can be isolated from the bones of *T. rex* and even older dinosaurs, some over a hundred million years old. These sequences of ancient proteins help us document, albeit indirectly, mutations that occur in DNA. They also augment classical morphology-based classification of long-extinct animals and plants.

But what about ancient biomolecules from really deep time? While we may never have DNA or even protein sequences from

300-million-year-old mollusks, corals, or crinoids, scientists have documented an amazing array of other kinds of ancient molecules: cellulose from plants, chitin from the exoskeletons of arthropods, and pigments as beautifully colored as those produced by organisms that live today. Our ability to identify ancient pigments, for example, has allowed us to reconstruct the color patterns of organisms such as feathered dinosaurs. These latter ancient biomolecules do not contain genetic information, but they still shed light on a wide range of questions about ancient functions and behaviors: If a 500-million-year-old organism produced a brilliant red pigment, does that mean that they—or perhaps their predators—could see and react to that pigment? When did color vision evolve? Was the evolution of skin and feather pigmentation involved in the evolution of sexual display and courting behavior?

In our examination of ancient biomolecules, we will travel back to the very origins of life, as well as to some of the most interesting places on Earth. We will travel to Yoho National Park in Canada, where we will examine the iconic organisms of the Burgess Shale; to the amber mines of the Dominican Republic, where we will find a very different environment than that depicted in *Jurassic Park*; and to Clarkia, Idaho, where we will split 15-million-year-old shale to expose leaves whose greenish color foretells the presence of the photosynthetic pigment chlorophyll. We will accompany scientists as they collect these fossils and follow them in the lab as they extract and characterize ancient biomolecules from fossils from deep time.

We will begin our journey by spending a bit of time with my blood-engorged mosquito, as a means of introducing the methods and materials involved in this new frontier of science. Then, in chapter 2, we will see just how far back in time ancient biomolecules are able to take us. The rest of the book is broadly organized by the different ancient biomolecules: in chapters 3 to 5, we will discuss ancient pigments, which help us understand the colors of ancient life, as well as the evolution of color vision. In chapter 6, we will turn our attention to ancient biometals. While you might not think of metals as biomolecules, they have shown wide application to the study of ancient life; chapters 7 and 8 tackle one of the most illuminating types of ancient biomolecules, proteins, which shed light on a wide range of topics—including evolutionary, behavioral, and physiological aspects of life in the past. The degree to which they extend into deep time also provides one of the more controversial topics in the field of paleobiology. After discussing proteins, we turn to perhaps the holy grail of ancient biomolecules: ancient DNA. Our ability to document changes in DNA through deep time has uncovered the genetic history of the evolution of many species, including, as we will discuss in detail, that of our own. Our discussion so far has primarily concerned the animal kingdom, but in chapter 11, we look at what we can learn about early plant life. We will learn of the amazing diversity of ancient plant biomolecules and about the field of chemotaxonomy to which that diversity has given rise. We will conclude our journey by turning our gaze from the past to the future: what new discoveries will the science of ancient biomolecules reveal? Will ancient genomes allow us to produce viable embryos and clone long-extinct animals? Will we be able to make proteins that existed billions of years ago? While I cannot, of course, provide a definitive answer to these questions, one thing is certain: by the time you finish this book, you will never think of a fossil in the same old way again.

I

A BLOOD-ENGORGED MOSQUITO

In northwestern Montana, there is an area known as the Kishenehn Formation. The name derives from the indigenous Kyunaxa peoples' name for white fir trees, Kishinena, which are plentiful at higher (7,000+ feet) elevations in the area. This is where I do my fieldwork. It is also where the blood-engorged mosquito met its fate and would be discovered millions of years later. The formation consists of rocks that formed from the accumulation of sediments in a large basin stretching all the way from Canada to the Bob Marshall Wilderness south of Glacier National Park. Created by a series of faults in the earth's crust, the basin slowly filled with water and became a huge lake. This lake is the primary reason that fossils exist in the formation.

Today, the basin holds both the North Fork of the Flathead River, as it descends from British Columbia, and the Middle Fork, as it flows northwards from its origin deep within the million acres of the "Bob." As the river exits the Wilderness near the small community of Essex, it begins to erode its way through 46-million-year-old shale rock derived from the shallow lake's sediments. The two forks of the river form the 60-mile-long western boundary of Glacier National Park, and it is along the rocky edges of the Middle Fork, with the park to the east and

Flathead National Forest to the west, where the thin shale contains arguably the best-preserved compression fossils of insects in the world.

At first glance, fossil insects may not be as sexy as *T. rex* or *Archaeopteryx*. But in terms of their diversity and sheer cumulative mass, insects are the most successful group of complex organisms on Earth. Scientists have named about a million species of living insects and estimate that there are many millions yet to be discovered. And then there are the millions of species of insects that are extinct, with a fossil record dating back nearly 400 million years.[1] If you want model organisms with which to study evolution, fossil insects are bursting with opportunities.

Unfortunately, the fossil record of insects is depauperate, a term oft used by one of my colleagues and one of my favorite scientific terms—it means lacking in diversity. There are only about 27,000 described species of fossil insects, far less than 1 percent of those that have ever existed. In the realm of flies (*Diptera*, or "two-winged" in Greek), scientists have identified over 150,000 living species, but only about 4,000 fossil species.[2]

When I first started my work on fossil insects, I was convinced that the older the fossil, the better. Insects have inhabited Earth for hundreds of millions of years; wouldn't it be cool to understand how they first evolved? Then I met Sonja Wedmann, head of the Paleoentomology Section at the world-famous Messel fossil site in Germany. Sonja reminded me that, despite their deep time[*],[3] origin, flies are still evolving. In fact, a huge radiation,

---

* The term "deep time" was coined by John McPhee in 1981, in an attempt to convey the unfathomable vastness of time to nongeologists. Some geologists dislike the term; others prominently display it in our museums in the hope that people will stop and think about the concept, if only momentarily.

or diversification, of dipteran species occurred fairly recently, shortly after the demise of the dinosaurs. These flies, which evolved to include what we refer to as "maggots" as an essential part of their life cycle, today make up more than a third of all living fly species. A subset of these, a relatively young group called the calyptrates, includes the well-known house fly, tsetse flies (now resident in Africa but which lived in Colorado 34 million years ago), flesh flies, and the parasitic bot flies that tormented Teddy Roosevelt and his fellow adventurers when they explored the Amazon.

The calyptrates today consist of more than 22,000 species, about one in seven of all living flies. Unfortunately, they too have a miserable fossil record, with only about 50 species described.[4] But due to some auspicious timing, this may soon change. The calyptrates are thought to have radiated at the beginning of the middle Eocene, around the time that the lake sediments that eventually formed the Kishenehn Formation were deposited. Hopefully, the fossil flies of the Kishenehn Formation will shed some light on this poorly documented radiation. Over the past 14 years, my colleagues and I have described new fossil species from nearly 30 different families. While relatively rare, the calyptrates are there too, waiting to be described.

Our aim in this work is not merely to provide taxonomic descriptions of new species. Rather, we hope to explain, among other things, why this explosive diversification—the evolution of thousands of new species—took place. About 50 years ago, Niles Eldredge and Stephen Jay Gould proposed the theory of punctuated equilibrium to explain such events: species can exist for very long periods of time, even geological time, until there is a sudden environmental change to which organisms must adapt. A beautiful example of this phenomenon occurred with the sudden cooling of the earth about 34 million years ago. The resultant replacement of tropical forests with vast temperate grasslands covering a million square miles of the Great Plains of North America is thought to have been a major factor in the evolution of the modern horse. What changed 45 million or so years ago that stimulated the rapid evolution of the calyptrate flies? Answering this question will keep me and many other scientists busy for many years.

Recently, however, the discovery of the fossil of the blood-engorged mosquito caused me to expand my work to the study of ancient biomolecules. The study of original molecular components—the pigments, blood molecules, and other biomolecules—that made up the insects flying around the subtropical shore of ancient Lake Kishenehn is the most fascinating work I have ever done, as it promises to produce enormous amounts of knowledge about life in deep time. Indeed, the Kishenehn Formation may eventually be better known for the biomolecules preserved within its fossils than its documentation of insect diversity.

Ancient biomolecules are, in one sense, laughably common. Atmospheric oxygen, which appeared roughly 2.4 billion years ago, is an ancient biomolecule, as are the trillions of tons of fossil fuels composed of degraded plant material.[5] But the ancient biomolecules found in the Kishenehn Formation and a few other sites around the world are rare because they are preserved in situ; that is, in their original position within a fossil with the anatomical context intact. This is important because it allows us to associate a structure and its ancient biomolecules with a function. To explain how biomolecules come to be preserved in situ, and how we can identify them, let's start with the discovery of the blood-engorged mosquito.

Discovery of the Kishenehn Formation's Ancient Biomolecules

Hiking in Glacier National Park is not for those who prefer to do their downhill skiing in Kansas. You're in the Rocky Mountains, so almost every trail includes dramatic climbs. As one rafts down the Flathead, the tectonic forces that shaped the park stare you in the face: the shale and siltstone slant up along the shores of the river at an angle of about 40 degrees. Exploring the river for fossiliferous exposures requires climbing on steep slopes composed of loose scree—with no one around to pick you up when you fall. (As the result of one particularly memorable tumble in 2020, I tore both rotator cuff and biceps tendons.)

Exposures of the Kishenehn Formation's shale rocks are most easily accessed by raft, but the put-in and take-out sites are so far apart that you can spend most of the day simply traveling to and from the fossil sites. The only other option is a long bushwhack through the Flathead National Forest. Because exposures of shale are present on both sides of the river, the fast-moving and very cold Flathead must be forded, both in the morning and in the late afternoon when your backpack is full of fossils. Unlike many of the lakes and streams in the park, which are filled with suspended glacial rock powder and milky in color, the Flathead is clear to the point where estimates of its depth can be seriously inaccurate. Unbuckle your backpack, use a hiking pole, and, given that the rocks at the bottom of the river are covered with a slippery layer of algae, wear water shoes with a good grip. And don't assume that the hole in front of you is two feet deep. A misstep that takes you into a five-foot-deep hole has several immediate effects: your lungs shrink to the size of a pea; your brain immediately screams "Panic!"; and the increased surface area of your body that is underwater drastically increases

FIGURE 1.1. One of several collecting sites on the Middle Fork of the Flathead River. Note the steeply angled banks of Kishenehn Formation shale.

the river's ability to push you downstream; as you pivot to turn back, that force increases even more. It took two summers of fieldwork to find two reliable places to ford the river.

Initial explorations of a seven-mile-long segment of the river in 2009 and 2010 revealed numerous but discontinuous exposures of fossiliferous shale. One stretch of the Middle Fork turned out to be one of the best places in the world to find compression fossils of mosquitoes: nearly a hundred different specimens have been found, including several that have been identified to the genus *Culiseta*, an interesting group whose living species feed on modern-day dinosaurs (which we normally call birds).[6] While there are well over 3,000 different living species of mosquitoes, and an unknown number of extinct species from deep time, only 20 species of fossil mosquitoes have been identified, and only five species are older than those found in the Kishenehn Formation.[7] Specimens have been found in Cretaceous amber, which makes them significantly older than those in the Kishenehn Formation, but they have rather short proboscises, and it is hard to tell if they were ancestors or simply cousins of today's mosquitoes.

In the fall of 2012, while photographing newly acquired specimens, I came across a fossil of what looked to be a blood-engorged mosquito. Most mosquitoes feed on blood because it is required for the development of their eggs. How and when did this lifestyle evolve? There is no evidence that the older mosquito fossils with short proboscises were hematophagic (blood-feeders). Even a long proboscis does not mean that a mosquito necessarily fed on blood. Both males and females of *Toxorhynchites theobaldi* have very long proboscises but feed exclusively on flower nectar—as do the males of all mosquito species. So how can we prove, beyond doubt, that a mosquito was hematophagic based solely on its fossil?

As a female mosquito's abdomen fills up with blood, she begins to excrete clear yellowish blood plasma, thereby making room for the more protein-rich red blood cells. Human blood is normally 40–45 percent plasma, so this strategy allows the female mosquito to load up on the blood cell protein required for egg production. That protein, which comprises 96 percent of the dry weight of red blood cells, is hemoglobin. It is red in color, binds oxygen, and because it does so reversibly, it can distribute oxygen throughout our body. Hemoglobin encounters and binds oxygen in the blood vessels of the lung, but in the small blood vessels of working muscle tissue, glucose is metabolized, and carbon dioxide is released; the CO_2 acidifies the blood and causes release of oxygen from hemoglobin.

The fossil of the blood-engorged mosquito certainly looked the part, with its distended abdomen and dark brown color. But appearances can be deceiving. We needed data, definitive data, that would confirm our hunch. If it was engorged with blood, it would be the first proof of hematophagy by mosquitoes, the single most economically disruptive fly that exists.

A major advantage of working at the National Museum of Natural History in Washington, DC, is access to the expertise of hundreds of colleagues. Conveniently, the Department of Mineral Sciences is just one floor above the paleobiology laboratory on the third floor of the Museum. Tim Rose, the manager of Analytical Laboratories in the department, is an expert in an X-ray spectroscopic technique that allows for the identification of the individual elements that make up a particular sample. I approached Tim about our specimen, and he enthusiastically agreed to perform a series of analyses on the various parts of the fossil. When examined in an electron microscope at a thousandfold magnification, the surface of our mosquito's abdomen had the appearance of a thick and seemingly

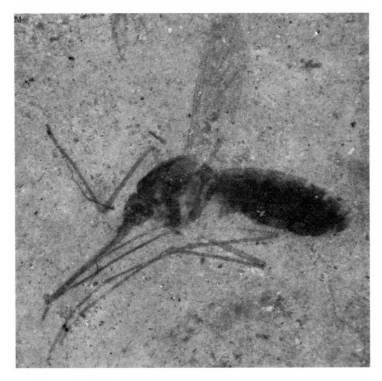

FIGURE 1.2. The first of numerous blood-engorged mosquito fossils discovered in the Kishenehn Formation. The abdomen contains remnants of blood from the insect's host.

homogeneous layer that resembled the smooth surface of the graphite core of a pencil. But what was it made of?

Tim's data showed that 67 percent of the weight of the blood-engorged female's abdomen was carbon. This was a good sign because living mosquitoes have been shown to have a very high carbon content (about 2.5 times that of our bodies). Analysis of the shale matrix adjacent to the fossil gave a value of 21 percent carbon—this was to be expected, as many of the crystalline minerals in the matrix, such as the carbonates, also contain carbon. These data suggested that the specimen's carbon originated from the body of the insect itself and was organic in nature. The fossil had not been permineralized—that is, its original biomolecular components had not been replaced by minerals. Might the carbon matrix of the fossil contain ancient biomolecules?

The hemoglobin protein binds oxygen indirectly via an intermediate, a flat ring-shaped molecule called heme. In the middle of each heme molecule, there is a single molecule of the metal iron, an ingredient required for the binding of oxygen. It is, in fact, the heme that is red and provides the larger molecule hemoglobin with its color. Heme is thus a pigment, a fact not lost on the vegetable-based hamburger manufacturers, who add heme to the mix to give the vegan "meat" its color. For us, heme's iron was an obvious target for analysis as it is easy to detect. If it had been preserved, Tim would be able to find it. Indeed, Tim demonstrated that the iron content of the blood-engorged mosquito's abdomen was nearly tenfold higher than that in the abdomen of a specimen that had not taken a blood meal, the fossil of a nonfeeding male. The huge concentration of iron was the smoking gun. I was ecstatic. The chemical evidence supported what our visual examination had led us to believe—that the mosquito's abdomen contained remnants of its host's blood.

To answer our next question—whether the entire heme molecule itself had been preserved—we required a different instrument, a mass spectrometer. The mass spectrometry technique is a bit like throwing a hand grenade into a foxhole and analyzing what comes out. The instrument bombards the surface of a fossil with heavy atoms of bismuth, the heaviest of the nonradioactive metals. As a result, fragments of ancient molecules are thrown up from the surface of the specimen and then whisked off to have the mass of each individual fragment recorded. The result is a unique pattern of fragments that provide a "fingerprint" that is unique to the original intact parent molecule.

In this case, the mass spectrometry technique returned the unique fragmentation pattern characteristic of heme. Numerous subsequent analyses demonstrated that the pattern was essentially identical to a positive control, purified pig hemoglobin. Moreover, the heme pattern was only present in the abdomen of the blood-engorged mosquito; it was absent from the abdomen of a male mosquito fossil collected from the same site, and absent as well from the rock matrix adjacent to the female mosquito. The evidence could not have been clearer: the mosquito contained remnants of the blood of an animal from which it took a blood meal. We had definitive proof that mosquitoes evolved a hematophagic lifestyle at least 46 million years ago. When did it first appear? Probably millions of years earlier, but proof will have to wait for the discovery of an older fossil of a blood-engorged mosquito.

Talk about an evolutionarily successful adaptation. It has been calculated that nearly 20 quadrillion mosquitoes are alive today, their cumulative mass more than a hundred billion pounds. They draw blood meals from a broad array of animals, even cold-blooded snakes. On what kind of animal did our blood-engorged mosquito feed? Although this mosquito could not be identified to a specific genus, other fossil mosquitoes collected in the area have been identified by Ralph Harbach, an expert in mosquito taxonomy at the Natural History Museum in London, to the genus *Culiseta*. As mentioned above, this group of mosquitoes feeds on a variety of hosts, including birds. And we know that there were birds along the shore of Lake Kishenehn because we have found the imprint of a small bird's foot in the shale. Was the blood in the abdomen of the fossil mosquito from a bird? Some questions will never be answered.

Given that an ancient biomolecule was preserved in the blood-engorged mosquito, the fact that it was heme was not surprising. Heme and closely related molecules are ubiquitous, essential to almost all forms of life, exceedingly stable, and have an impressive fossil record. Heme has been shown to be preserved in fossils that are over 300 million years old, and "heme iron" has been reported from several different species of dinosaurs, including *Tyrannosaurus rex*.[8] Biomolecules related to heme go back much, much further. The light-harvesting pigment chlorophyll has a heme-like molecule at its center (containing magnesium instead of iron), and it is as old as the earliest photosynthetic bacteria—reaching back billions of years. Because of heme's stability, and the fact that it is not known to be synthesized abiotically (in the absence of life), it has even been proposed as a target in NASA's search for life on Mars.

When the rover Perseverance recently landed, it carried two instruments, PIXL and SHERLOC, that can analyze the rocks of Mars using several different analytical techniques, including X-ray spectroscopy. They can perform elemental analysis and detect the presence of iron. They can also detect organic compounds and, more importantly, provide "fingerprint"-like data that can be used to establish the molecular structure of those compounds. If they find heme, it will radically change our understanding of the universe.

What Is a Fossil?

The preceding discussion has focused on the surprising discovery of ancient biomolecular material inside of a fossil. But what, exactly, is a fossil? We often think of fossils as being made of rock: bones are buried and slowly become stone. This occurs because the wet sediments in which a dead organism is buried are chock full of dissolved minerals just waiting for a place to crystallize. As the original organic components of the dead organism decay, the resulting empty spaces are filled with crystallizing minerals. Eventually, no traces of the original animal remain. The process, called permineralization, is very slow, exceedingly precise, and can mimic morphological forms with a very high degree of fidelity. Many fossils are works of art.

Surprisingly, permineralization has been shown to begin quite rapidly after burial. Derek Briggs, a Professor and Curator at the Yale Peabody Museum of Natural History and an expert in the processes of fossilization, has shown that twigs buried in marine-like sediments start to become mineralized in as little as 80 days; other researchers have demonstrated that permineralization of cancellous (spongy) bone buried in simu-lated river sand and flooded with ion-rich fluid begins within 12 weeks.[9] At what point do these specimens become fossils? When they are 100 percent mineralized? Or 50 percent miner-alized? Does a 500-million-year-old sponge from the famous Burgess Shale in the Canadian Rockies, which contains small amounts of a component of its original exoskeleton, not qualify as a fossil because it is only 99.9 percent permineralized? How about partially permineralized 10,000-year-old Neandertals or 1,000-year-old medieval bones? Answers to these questions will necessarily be subjective.

There are also many types of fossils that are preserved with-out permineralization. Insects entombed in "fossilized" tree resins are invariably referred to as fossils in the scientific litera-ture and constitute an essential part of the fossil record. For example, insects in 120-million-year-old amber from Lebanon and 99-million-year-old amber from Myanmar are some of the most spectacularly preserved and scientifically important fossil insects we have. There are also fossils preserved in what one might call more interesting circumstances, such as happens with pack rat middens. The 20 or so species of the genus

Neotoma are small rodents that have a habit of collecting bones, seeds, dead insects, etc. and bringing them back to their nests, on which they routinely pee. Substances in the urine crystallize and glue the nest together. In arid southwestern North America, the nests are often built within caves and can last for as long as 50,000 years in the absence of any form of permineralization. Full of fossils of a wide array of organisms, the nests of middens are an invaluable resource for paleobiologists and anthropologists. Yet another example to ponder: the organisms, from saber-tooth tigers to beetles, preserved in the La Brea tar pits in Los Angeles. Fossils, yes; permineralization, no. Despite common definitions that require a fossil to be made of rock, of a minimum age, and naturally produced (ruling out embalmed mummies), I prefer the definition provided by Britannica Kids: "Fossils are the remains or traces of plants and animals that lived long ago."

To survive the journey from the distant past to our present, every fossil needs a bit of luck. Some, however, require a tremendous amount. My blood-engorged mosquito, for example, shouldn't have had a snowball's chance in Hades of becoming a fossil. If it had fallen onto water, choppy waves or currents may have quickly popped the balloon and fragmented the insect. As it sank to the lake's bottom, chances were good that it would have been eaten by a fish or other predator. Once on the surface of the bottom mud, it was a sitting duck for predaceous worms and crustaceans. Lastly, it had to survive compression by subsequent lake sediments without being destroyed. What were the chances? In the foreword to the issue of the *Proceedings of the National Academy of Sciences* in which we described this 46-million-year-old mosquito, Derek Briggs observed that such fossils "force us to revise our ideas about the limits of fossilization."[10]

The preservation of the mosquito's original organic biomolecules was even more improbable. Our analyses of the blood-engorged mosquito showed that it had not been permineralized. Its body had been severely flattened, but it was still carbonaceous—much of its original carbon content had been preserved.[11] How could this have happened? As it turns out, evidence suggests that the mosquito was preserved by a process much like that by which the microbes that form stromatolites are preserved.[12] The mosquito fell not onto the water's surface but onto the surface of an algal bloom—imagine green slime— that happened to cover the lake. The sticky surface of the algae trapped the insect, and as the microbes continued to grow, they enveloped the mosquito and protected and preserved the fragile insect, even after the mat sank to the lake's bottom.

Improbable as the preservation of remnants of hemoglobin in the fossil of a blood-engorged mosquito is, even more improbable is preservation of a cholesterol-like molecule in a 380-million-year-old crustacean, or the preservation of chitin, the primary component of the arthropod exoskeleton, in a fossil from the 505-million-year-old Burgess Shale. These are only a few of many spectacular examples of ancient biomolecules that we will encounter in the following chapters. As more young scientists join this new area of research and modern analytical techniques become increasingly sensitive, the numbers and diversity of ancient biomolecules will surely increase, as will the age of the fossils in which they are preserved. How old might ancient biomolecules be? If we hope that Perseverance might find ancient biomolecules on the surface of Mars, which is thought to be between 3.9 and 4.6 billion year old, should we not also hope to find ancient biomolecules of a similar age here on Earth? To begin to answer this question, let's start with the oldest of Earth's fossils.

2

IN SITU

The mountains of Glacier National Park are the reason that three million people each year make the 50-mile drive over the Going-to-the-Sun Road. The mountains were sliced and diced by glaciers to produce astoundingly steep cliffs that reveal uncountable multicolored layers of ancient sediments. The 1.5-billion-year-old Appekunny mudstones formed by these sediments are readily visible from the road, their identification made easy by their distinctive green color. Many of the large pieces of mudstone that have fallen from the cliffs' higher reaches have surfaces with impressions left by ancient ripples and cracks that developed when a muddy beach dried out, indications of an ocean's presence.

In many parts of the park, one can also see large limestone cliffs, remnants of billion-year-old reefs that were made by ancient cyanobacteria. Often erroneously referred to as blue-green algae (cyanobacteria are blue-green in color, but they are not algae), the bacteria formed structures called stromatolites, which are essentially a series of discrete layers of bacteria and thin films of sediment. Over time, layer by layer, ancient reefs were created. Some forms of stromatolite are spherical and, when eroded, resemble the flat surface of a cabbage cut in half.

FIGURE 2.1. An ancient stromatolite reef adjacent to the Going-to-the-Sun Road in Glacier National Park. At the right, a single fossil stromatolite, approximately two feet wide, at a second site on the same road; despite the presence of a turn-out at the latter site, nearly all the Park's visitors remain unaware of this beautiful outcrop's existence.

Once you see the first one, the second is easy to recognize. Cliffs that border much of the most popular hike in the Park, the Hidden Lake Trail, are full of their fossils, and blocks of stromatolite rock as big as a car can be found on the slopes of Bear Hat Mountain, which towers high above Hidden Lake's western shore. These reefs are so common that anyone looking will easily find them—unfortunately, however, they go unseen by more than 99 percent of the park's visitors.

Cyanobacteria, which may have produced the first stromatolites several billion years ago, remained the dominant life-form on Earth for much of its early history. These bacteria were photosynthetic and used several different pigments, including chlorophyll, to capture the energy of the sun and produce oxygen as a waste product. Eventually, starting about 2.4 billion years ago, they were responsible for what is called the Great Oxygenation Event, the very gradual increase of atmospheric oxygen from essentially zero to something resembling today's level of about 21 percent. Newly evolving animals made the sudden availability of oxygen a centerpiece of their metabolism. When they died, they left fossils for us to find. And within those fossils, the possibility of ancient biomolecules.

Following the Fossil Record

How far back in time do the oldest fossils reach, and which of these fossils contain ancient biomolecules? In this chapter, we explore these questions by taking a brief tour through the earliest stages of life on this planet, as told by fossils. We will eventually arrive at one site of particular interest, the Burgess Shale in British Columbia, where ancient biomolecules have been found in (and have helped identify) half-billion-year-old sponges.

The origin of life on Earth is unquestionably one of the most interesting of the many unknowns in modern science. It is also one of the most difficult, if not impossible, questions to address. Most paleobiologists work with life's origins at more tractable levels: the first segmented animal, the first vertebrate, the first fish to walk on land. But long before there were living organisms, there were chemicals in primordial chemical soups that evolutionary biologists refer to as "WLPs" or warm little ponds. Cycles of alternating dry and wet conditions may have concentrated chemicals together; minerals in clay at the pond's bottom may have bound a subset of the chemicals together and allowed for the assembly of more complex molecules. Or perhaps lightning was involved. Somehow, after an uncountable number of random associations, the many exceedingly complex chemicals characteristic of life were produced.

Included among these were the components of deoxyribonucleic acid (DNA). Most people know, to some extent, that DNA encodes the sequences of the 20,000 or so proteins of the human body. But what most people don't know, and what really boggles the mind, is that DNA can't be replicated, can't contribute to the synthesis of proteins, and can't be repaired or properly degraded without the help of proteins. But how, then, did the first DNA molecule make the first protein in the absence of proteins? Might amino acids, the subunits of proteins, have self-assembled, without the need for DNA? We don't and may never know; the earliest precursors of life's biomolecules exist only in our imagination.

Life itself is thought to have originated on Earth surprisingly early in its history. The earth formed about 4.5 billion years ago, and there is evidence, albeit controversial, of life in rocks that are about 4.2 billion years old.[1] The data, from rocks found in Labrador, Canada, involve ratios of isotopes of the element carbon in tiny grains of graphite that suggest the existence of metabolism characteristic of life. Setting aside this putative evidence for the earliest of Earth's life-forms, there have been a number of reports of fossil bacteria that are billions of years old. However, definitive identification of a microscopic bacterium in rock is a tough assignment, and many such reports remain controversial. The first widely accepted structures assigned to bacteria are the stromatolites. The oldest, found in 3.7-billion-year-old rocks adjacent to the icecap in southwestern Greenland, are, at most, two inches in height, yet they exhibit clear evidence of the layering characteristic of stromatolites.[2]

The window between this bacterium and the oldest fossil of a more advanced multicellular life-form is huge, about 1.6 billion years. These multicellular organisms, much like the cyanobacteria that build stromatolites, form large entities in which every organism is identical to its neighbor. The appearance of Metazoa, truly multicellular organisms composed of two or more morphologically and functionally different cells (e.g., muscle and nerve cells), was perhaps the most important evolutionary event in the history of life. And according to the fossil record, it occurred at least 2.1 billion years ago.

In 2014, Abderrazak el Albani and his colleagues discovered what are referred to as the Francevillian biota, near Franceville in the southeastern corner of the Republic of Gabon.[3] Embedded in 2.1-billion-year-old black shales of the region are fossils of several different and distinctly complex organisms. They are large, approximately one centimeter in length and diameter, and display wondrously varied morphologies. There are lobed, elongate rod-shaped, and disk-shaped forms. Some of the organisms appear to have an anterior-posterior axis, perhaps distinct heads and tails. There is no other billion-plus-year-old locality with such a variety of complex life-forms. Many scientists think that the appearance of these organisms was directly related to the initial stages of the Great Oxygenation Event, when atmospheric oxygen levels, albeit very slowly, began to increase.

There have been other credible reports of billion-year-old complex life, including fossils identified as algae in a 1.9-billion-year-old deposit in Michigan and in 1.2-billion-year-old rocks in the Canadian Arctic Archipelago.[4] Yet another organism, of particular interest because of its size, is an unnamed specimen from northern China. Found in 1.6-billion-year-old mudstone, it is exceedingly large, about a foot long and as much as several inches wide.[5] Lingulate in shape, it has longitudinal striations and holdfasts. Examination of its surface reveals an extraordinarily unique meshwork of microscopic cells about 1/1,000 of an inch in diameter. Another unusual characteristic of these organisms is their preservation as unusually thick black carbonaceous compressions. Unlike the fossil impressions with which most people are familiar that resemble the imprint that a leaf produces when it lands on the surface of a freshly poured concrete sidewalk, these fossils are not impressions at all—instead, they are the compressed remains of the organism. Has the black material been degraded to pure carbon, like graphite, or does it contain intact biomolecules (reminiscent of the abdomen of the blood-engorged mosquito from the Kishenehn Formation)? No one, to date, has performed a chemical analysis of the remnants of this fossil.

The difficulty of interpreting fossils of a billion-year-old metazoan is exemplified by *Horodyskia*, one of Earth's more contentious early complex organisms. These fossils, which originate within the 1.5-billion-year-old Appekunny mudstones of Glacier National Park, were first collected by Robert Horodyski in 1982.[6] The fossils look like "beads on a string," with beads as large as 5 millimeters in diameter, while the entire organism is as long as 70 millimeters. Horodyski himself wasn't quite sure what he had discovered and suggested that they were "dubiofossils," geological in origin. Over the next 30 years, different scientists identified them as "pseudofossils," bacteria, algae, a sponge, slime mold, and even feces.

In 2013, Gregory Retallack and his colleagues visited Glacier National Park and collected *Horodyskia* from several sites along its eastern border.[7] Their microscopic examination of very thin slices of the fossils revealed that the oval "bead" structures were connected by a system of radial filaments that themselves possessed an internal structure. Retallack suggested that *Horodyskia*, which has now been found at several localities around the world, were sessile (immobile) organisms that lived in a shallow and warm, perhaps subtropical, sea. He offered up lichen as one possibility. However, the story has recently taken another turn with work by Roy Rule and Brian Pratt of the University of Saskatchewan in Saskatoon, Canada.[8] They have suggested that the beads of *Horodyskia* are mud particles trapped on filamentous tufts of a microbial mat. Is *Horodyskia* a

FIGURE 2.2. A photograph of what may be 1.5-billion-year-old organisms from Glacier National Park, collected by Robert Horodyski and later named *Horodyskia*. Scale bar = 3 mm.

metazoan or a "microbially induced sedimentary structure"? The debate remains unsettled.

The elapse of another billion years or so finds us in a period called the Ediacaran, named after the Ediacara Hills, a mountain range just north of Adelaide, Australia. The name itself is an aboriginal term for a narrow spring. Fossils from this era were first found in Newfoundland in 1868, but for nearly a century they were considered to be artifacts, based largely on the scientific establishment's belief that the younger Cambrian period (542–485 million years ago) was the cradle of complex life. During the last few decades of the twentieth century, however, Ediacara fossils were found at sites around the world. The organisms are numerous and diverse, with over 200 different genera described. Gradually, paleobiologists realized that nature—understood here as simply the accumulation and subsequent effects of random genetic mutations over time—was offering up a new window on the evolution of life.

The most common term applied to these organisms is enigmatic. The oldest date to about 600 million years ago, and the impressions they left in the rocks looked like nothing ever seen before.[9] Many of the fossils look like the rubber floor mats that we use in our cars; some are disk-shaped, some oval. Some show signs of being bilateral, even triradial—but they rarely exhibit any indication of a head, mouth, or anus. Were they plants, animals, or something altogether new to science? Nearly all species from this era eventually disappeared from the fossil record, and only very few even made it as far as the Cambrian. They were not, however, the only complex life on the ocean floor at that time. Also present were sponges, a few arthropod-like and echinoderm-like organisms, and a bizarre organism called *Cloudina*, which lived in a long tube of nested vase-shaped structures made of calcium carbonate. *Cloudina* has recently been shown to have a digestive tract with a distinct mouth and anus—it was an animal.[10]

Biomarkers and Biomolecules in Situ

To what extent can our history of deep time fossils be overlayed with a corresponding record of ancient biomolecules? The answer depends on whether we require the biomolecule to be

present in a physically recognizable fossil, in other words, preserved in situ (Latin for "in its original place"). For example, the
heme found in the abdomen of a blood-engorged mosquito,
exactly where we would expect it to be, was preserved in situ.
Its presence possesses physiological and behavioral context. On
the other hand, if heme were extracted from the powder of a
handful of crushed rock, the finding has no context. Given that
heme can't be made by purely geological (abiotic) reactions, its
presence in the rock may tell us that it once harbored life—but
what form of life? Ancient biomolecules that are not preserved
in situ are called biomarkers.

Although they are no longer physically associated with fossils of the organisms of which they were once a part, biomarkers
can still be of great value, since many biomolecules are known
to be specific to certain types of life. Photosynthetic bacteria,
algae, complex plants, and even fungi that degrade wood are a
few examples of organisms that can be identified by organism-
specific biomarkers. One particular biomarker has been used to
document the existence of 1.6-billion-year-old bacteria from
Australia that metabolized sulfur instead of oxygen.

For many years, the oldest evidence for the existence of
sponges was a biomarker, a sponge-specific derivative of cholesterol.[11] Found in 650-million-year-old rocks in the Arabian
Peninsula, this biomarker demonstrates the potential of ancient
biomolecules to provide us with taxonomic information about
very ancient organisms and the evolution of life on Earth, even
if they are not found in situ. But never underestimate the fossil
record. Recently, Elizabeth Turner has reported the presence of
three-dimensional branching tubules in rocks from the
890-million-year-old Little Dal reef, not far from the Mackenzie
River in northwestern Canada.[12] Turner did not, however, ven-
ture to give the specimens a genus or species name. Rather, she

simply referred to them as "vermiform microstructures" and guardedly stated that "If the Little Dal objects are truly sponge body fossils," then they are the oldest known representatives of that phylum.

Two of the more interesting uses of biomarkers involve the characterization of the environment following the extinction of dinosaurs. Sixty-six million years ago, the Chicxulub asteroid, about 7.5 miles in diameter and travelling at nearly 12.5 miles/second, collided with what is now the Yucatán Peninsula in Mexico. With an energy calculated to be several billion times greater than that of the nuclear bomb dropped on Hiroshima, the collision caused what is known as an "impact winter," a dramatic cooling of the world's climate. Large increases in atmospheric dust and sulfur, along with smoke from forest fires that blanketed large portions of the earth, blocked the sun. Average global temperatures plummeted nearly 13 degrees Fahrenheit. How do we know it was 13 degrees? As it turns out, there is a cliff on the Brazos River in Texas that contains layers of sediment laid down just after the impact—sediment that contains unique biomarkers. These biomarkers are lipids found in the external membranes of certain types of ancient microbes. Johan Vellekoop and his team knew that living species of these organisms change the composition of the lipids (fats, oils, etc., insoluble in water) of their membranes in a temperature-dependent fashion, and so they were able to use lipid composition of the Brazos River cliff microbes as if it were a 66-million-year-old thermometer fixed in time.[13]

Equally fascinating are studies reported by Bettina Schaefer. Knowing that many biomarkers are organism-specific, Schaefer and her colleagues wondered if biomarkers could inform us about the post-impact recovery of life. Sampling an eight-inch-diameter core that a team of international scientists had drilled in the still-existing ring of the Chicxulub crater, Schaefer and her colleagues were able to identify layers of sedimentary rock that had been deposited immediately after the impact.[14] Schaefer's team then started to look for biomarkers. What they found allowed them to write the script for a time-lapse video that depicted more than a million years of post-impact recovery. The bottommost biomarkers were unique to organisms that were characteristic of coastal areas, including plants and the cyanobacteria that built stromatolites. There was even a biomarker that provided evidence of combustion, the existence of which was confirmed by the presence of charcoal in the sediments. The researchers were able to show that the asteroid impact had spawned a massive tsunami that reached and swamped coastal lands. When the wave receded, it carried plants, stromatolites, and wood that had been scorched by the heat of the impact. This material, along with their specific biomolecules, eventually sank to the bottom of the crater and were incorporated into the sediment. The researchers went on to document massive and sustained post-tsunami blooms of green-slime-producing blue-green algae. With the crater full of nutrient-rich detritus, the slime covered the area for about 200,000 years. As the darkness and cold of the impact winter slowly waned, other biomarkers in the sediments documented the appearance of different types of photosynthetic organisms. It took North America millions of years to fully recover.

As these examples demonstrate, biomarkers can serve as valuable sources of information about the distant past. Yet biomarkers are rarely the original biomolecules that functioned in living organisms; most have been fragmented, degraded, condensed, cyclized, and/or polymerized. Ancient biomolecules preserved in situ provide a different, perhaps somewhat clearer window into the past. Our ability to identify such biomolecules is, however, still in its infancy. To begin our investigation into the different kinds of ancient biomolecules preserved in situ, let's start with the oldest: cholesterol.

Cholesterol is a biomolecule that most people know because of its potential effect on our health—it has a nasty habit of contributing to the formation of plaques in our blood vessels. But this essential biomolecule, the chemical structure of which resembles a small piece of chicken wire with several intact six-sided holes, has been around for over a billion years. Since then, evolution has tweaked that structure to produce a large repertoire of related biomolecules that have an amazingly diverse array of functions. Cholesterol itself regulates the fluidity of our cell membranes—the higher the percentage of cholesterol, the less fluid the membrane. Plants don't make cholesterol, but they do make the structurally related phytosterols (the Greek word *phyto* means "plant"), which serve the same function. Bacteria do not make cholesterol.

The presence of bacterial-specific analogues of cholesterol, however, can be used to identify traces of ancient cyanobacteria. Other biomolecules related to cholesterol allow us to differentiate animals from plants. Recently, scientists from Australia have done just that, providing evidence for the oldest ancient biomolecules identified in situ. From 2015 through 2017, Ilya Bobrovskiy and Jochen Brocks travelled to the White Sea Region of northeastern Russia, where they collected numerous specimens of the Ediacara organism *Dickinsonia*, a genus that could easily be the poster child for the Ediacara biota. Up to four and a half feet long, with the appearance of a giant oval hand-woven fan, these 558-million-year-old specimens were unique because they were covered by a thin black layer of organic matter. Although this thin organic film was less than 1/10,000th of an inch thick, when extracted and analyzed for ancient biomolecules, it was found to contain the carbon skeletons of biomolecules related to cholesterol that were clearly associated with the fossil. Extracts of the sandstone immediately below and above the fossil had nearly tenfold fewer of the cholesterol-related molecules than the fossil itself. And, unlike extracts of the fossil, extracts of the sediment-derived matrix were dominated by different biomarker molecules, which were characteristic of green algae that lived in the ancient sea. The cholesterol-related biomolecules isolated from the fossil were stable degradation products of the cholesterol that had once been part of the organism, clear evidence that the Ediacaran genus *Dickinsonia* was a multicellular animal.[15]

Cholesterol-like biomolecules also played an important role in a discovery made at Mazon Creek, one of the most scientifically valuable fossil localities in the world. Located about 60 miles southeast of Chicago, Mazon Creek's several sites are renowned for the preservation of soft tissues, and more than 700 extinct species of plants and animals have been found there. Although most of the animals are invertebrates, perhaps the most striking fossils ever found at the site are those of intact juvenile sharks, the longest of which is about six inches long. The 309-million-year-old Mazon Creek fossils are unusual in that they occur within what are called concretions. Shaped like slightly flattened eggs, concretions range in size from an inch to a foot in length; they are composed of sand and clay, which have been cemented together by iron carbonate, a mineral known as siderite.

In early 2020, Kliti Grice, the Director of the Organic Geochemistry Center at Curtin University in Perth, Australia, toured several major museums and universities in the United States in search of such concretions. When she met with curators at the Natural History Museum in Washington, DC, she hit the jackpot. Not only were they able to provide her with numerous beautiful Mazon Creek specimens of thick intact ferns nested within split concretions, they also sent her back to Australia with a huge sack of unsplit concretions.

FIGURE 2.3. A 307-million-year-old Mazon Creek nodule (or concretion). The split nodule contains a fossil of the sea cucumber genus *Achistrum*.

Dr. Grice's interest in the Mazon Creek fossils stemmed from her earlier work in which she and her colleagues had isolated steroids and other biomolecules from concretions collected in North Western Australia. One particular specimen, a 380-million-year-old crustacean, was found to contain original cholesterol-like biomolecules—not degraded forms of these molecules, but the original biomolecules that functioned in the living animal.[16] This was a huge advance in the field of ancient biomolecule research. Grice's lab had shown that, under the right preservational conditions, complex labile biomolecules— identical to those that existed in an extinct organism when it was alive—could survive in situ for several hundred million years.

Ancient Arthropod Exoskeletons

Most fossils of multicellular marine organisms are preserved as the result of having a hard shell. Mollusks and brachiopods are good examples. Unfortunately, organisms without shells, which may have comprised the majority of life near ancient reefs, are rarely preserved. A wonderful exception, however, is the fauna of the Burgess Shale in British Columbia, which was discovered by accident in the 1880s. While building a railroad through the Canadian Rockies—or, as local climbers call them, the Canadian Softies, due to their soft shale, limestone, and mudstone composition—the surveyors found "stone bugs" high up on Mount Stephen. These trilobites were plentiful, and the local founding fathers advertised their existence in attempts to attract tourists to the area. Despite their local fame, it took nearly 20 years before Charles Walcott, a paleobiologist and the secretary of the Smithsonian Institution, heard about the fossils and decided to investigate. What Walcott found was not your typical fossil.

The Burgess Shale fossils appear as dark patterns on black shale. Mollusks and brachiopods are present, but most of the fossilized organisms were originally soft and without any type of shell—think of something resembling a worm, jellyfish, or sea anemone. The organisms are straight out of a science fiction writer's imagination, and their names, such as *Hallucigenia* and *Anomalocaris*, are indicative of the perceived strangeness of the organisms.[17] Indeed, many scientists criticized Walcott for concluding that the Burgess Shale's apparently new and unique fauna belonged to existing groups or taxa of animals, classifying many of them as arthropods. In the early 1970s Harry Whittington at Cambridge University and his two soon-to-be-famous graduate students, Derek Briggs and Simon Conway Morris, initiated a re-evaluation of the Burgess Shale fauna and concluded that many belonged to heretofore unknown taxa. New and more detailed specimens continue to be found; even new sites continue to be discovered. In 2012, for instance, scientists from the Royal Ontario Museum in Toronto published a description of the Marble Canyon site, only 26 miles from the original Walcott Quarry location. While the Marble Canyon fauna contains fossil species in common with the original site, it has also yielded numerous new species. Despite living 508 million years ago, these organisms had eyes, mandibles, and gills. Perhaps most amazing are *Pikaia gracilens* and *Mettaspriggina walcotti*, basal representatives of the phylum Chordata: these animals have a notochord,*,18 a primitive precursor of backbones.

If you are up for a 13-mile hike, you can reach the exact place where Walcott first discovered the fossils. You can stand on pieces of shale that may have been there when he looked higher up the ridge in an attempt to locate the source of the fossiliferous shale that had tumbled down over the trail. And you can hike the switchback to that source, now called the Walcott Quarry, to examine specimens of many of the nearly 150 species of fossil organisms that constitute the Burgess Shale fauna. The hike starts at a mile-high trailhead on the Yoho River, and when I was there in early August, the river was high, fast, and the color of dirty milk—British Columbia eroding its way down to the Pacific Ocean. From the 1,250-foot Takakkaw falls, the hike ascends half a mile over the course of about six miles. Views of Emerald Glacier, Emerald Lake, and numerous snow-capped mountain peaks help to take your mind off the sometimes-arduous nature of the hike.

Near the end of the hike, you meander through fields of large, nearly rectangular pieces of inch-thick rock. Although they may look like they would make a great slate roof, they are in fact pieces of shale, sediments of a 508-million-year-old sea bottom. At the quarry, thousands of pieces of rock are strewn over the ground,

---

* In humans, the notochord appears, transiently, in the embryo and plays an important role in the formation of the vertebral column. Chordates from China, such as *Haikouichthys*, are about 30 million years older than the Burgess Shale fauna and have actual backbones with clearly visible vertebrae.

many of which contain some of the most iconic fossils ever found. You can pick them up, photograph them, even make impression tracings of them, but, before you head back down with your guide, you must leave them where you found them. Alaina Petrella, the guide for my group of five, recounted a story about Emily Taylor, who, on a guided hike to the quarry, found a specimen that was subsequently described as a new species. Inspired by the story, the five of us bent down to turn over pieces of shale with renewed energy and enthusiasm. I brought a fossil to Alaina and asked for an ID. "It's *Vauxia*," she said. *Vauxia*, named for Mount Vaux in Yoho National Park, is one of the oldest fossils known to have preserved original biomolecular components. Buried by one of the periodic mudflows that cascaded off a reef some 200 miles from land, *Vauxia*, a sponge, became an improbable fossil; it has been shown to contain the biomolecule chitin, a structural component of its original fibrous skeleton.

Chitin is one of the most common polymeric biomolecules on Earth. Humans don't make it; in fact, no vertebrate does. However, it is a major component of the exoskeleton of a huge number of invertebrate organisms. The hard, external cuticle of arthropods, including the insects that comprise the majority of multicellular species on the planet, contains chitin. One gets an idea of how tough mineralized chitin is when attempting to crack open the leg of a Maryland Blue Crab. Chitin is a polymer of a derivative of the very common sugar glucose. Nature, it turns out, is an incredibly efficient chemist. Over half a billion years ago, it used variations on a single theme—polymers of glucose and its derivatives—to generate chitin, cellulose, glycogen, and starch, which serve a variety of functions in a wide range of organisms.

The exact structure of the chitin matrix is not well understood. Long chains (polymers) of the sugar are bound together to make microfibrils of chitin, which are embedded in a meshwork

FIGURE 2.4. Alan Munro, a graduate student at George Mason University, at the Walcott Quarry in 2019. The slab of shale he is holding contains several large specimens of the carnivorous worm *Ottoia prolifica*. Living in a U-shaped burrow with their head at one end and the body anchored in the other, this organism is related to today's penis worms. Fossils of brachiopods, organisms that are unrelated to but superficially resemble clams, have been found within the fossil worm, indicating that they were a favorite prey.

of proteins—in insects there are as many as a hundred such proteins, some of which are site-specific. Some provide hardness, while others allow the cuticle to stretch like rubber, handy for a cuticle that surrounds the base of a moving appendage, such as a wing. Most of these proteins, however, are poorly characterized. In fact, it wasn't long ago that all of these different proteins were collectively referred to as arthropodin, as if they were a single component.

When the cuticle of an arthropod hardens—for example, after each molt, during which the old exoskeleton of an insect nymph is discarded and a new one formed—the chitin microfibrils become extensively cross-linked, both to other chitin fibrils and to the many cuticular proteins. The cross-linked matrix is exceedingly complex, and even the most sophisticated modern spectroscopic techniques are unable to provide us with its exact structure. This is, in part, because there are many variations to the chitin theme. The hardened cross-linked molecule is also incredibly stable—even at a temperature of 360 degrees centigrade (680 degrees Fahrenheit), it does not easily degrade.[19] Given this extreme durability, it is perhaps not surprising that it is found in the fossil record, preserved for hundreds of millions of years.

When Hermann Ehrlich and his colleagues undertook the challenge of examining the composition of *Vauxia gracilenta*, they realized that no single experiment or technique would suffice. If they were to report something as startling as the preservation of a 508-million-year-old ancient biomolecule, they would need to present nearly definitive data. To meet this burden, they decided to use a half-dozen different techniques. First, they chopped it up to find defining chitin-specific fragments. They isolated chitin from living sponges as positive controls and demonstrated that such fragments were present in both *Vauxia gracilenta* and the living sponges.[20] Ehrlich and his colleagues also took advantage of the fact that nature has devised methods to degrade its garbage; if the exoskeletons of dead insects were never broken down and recycled, we would be neck-deep in dead flies and beetles. An enzyme purified from a fungus, called chitinase, which can cleave chitin, was shown by Ehrlich and his colleagues to destroy the chitin from *Vauxia gracilenta*. And importantly, there was no evidence of chitin in the rock matrix adjacent to the fossil—the chitin in *Vauxia*'s exoskeleton had been preserved in situ.

Chitin is one of the most common biomolecules on Earth and, as a result, ancient chitin has been found in a variety of deep time fossils. These include a 310-million-year-old fossil of a scorpion from Illinois, a 200-million-year-old snail egg case from Poland, a 34-million-year-old cuttlefish *Mississaepia mississippienis* from—you guessed it—Mississippi, and, not surprisingly, a relatively young 25-million-year-old beetle from Germany.[21] It has also been found in a 417-million-year-old euryptid *Eurypterus dekayi* (an extinct relative of scorpions, often referred to as a sea scorpion) from Ontario, Canada.[22] Some evidence suggests that the chitin in this specimen is still present as a chitin-protein complex. However, only hints of proteins, in the form of individual protein subunits (amino acids) and very short fragments, have been reported. As we will see in a later chapter, many scientists believe that the oldest known ancient protein is less than 4 million years old; 417-million-year-old protein sequences may be a bit of a stretch.

The chitin in the 508-million-year-old Burgess Shale sponge is of particular interest for reasons other than its age. The presence of chitin was critical to the classification of the animal. A quick and dirty classification of the sponges separates them into sponges with no skeleton (for example, slime sponges, which, not surprisingly, have no fossil record), and sponges with either a fibrous skeleton or skeletons made of minerals such as calcium carbonate or spicules of silica. Discrimination between living representatives of the latter two groups is straightforward, but in fossils, the task can be much more difficult. There is an old adage in paleobiology that goes "Absence of evidence is not evidence of absence." A fossil sponge without preserved spicules may never have had spicules, or the spicules may have been dissolved or otherwise lost over the passage of time. *Vauxia gracilenta* was given its genus name in 1920 by Charles Walcott, and Keith

Rigby of Brigham Young University subsequently suggested that it belonged to a group of very primitive sponges in the group Keratosa, which were thought to have fibrous skeletons.[23] However, the morphology of these fossil sponges is poor, and their classification understandably imperfect. The presence of chitin in this Burgess Shale fossil was a critical clue that confirmed *Vauxia* as a member of the Keratosa—Rigby was right. *Vauxia* provides us with another important insight: biomolecules other than those that provide sequence data (i.e., proteins and DNA) can provide important taxonomic information.

Preserved ancient biomolecules can even help us understand how organisms became fossils. When organic-rich material (leaves, algae, or animals) falls to the bottom of a lake and becomes deeply buried and heated, it is nearly always degraded to the point of being unrecognizable. The resulting material is called kerogen, an ill-defined, complex, cross-linked, and insoluble material that often degrades further into components of coal, oil, and natural gas. In extreme cases, the material degrades to the point where nothing but pure carbon, in the form of graphite, remains. But kerogen itself, if it retains anatomical features of a fossilized organism, can produce beautiful fossils in the absence of permineralization.

At an age of 508 million years, the fossils of the Burgess Shale were always assumed to be fossilized via the process of mineralization. But in 1990, Nicholas Butterfield, then a student at Harvard University, described a process he called "fossilization by way of organic preservation."[24] He demonstrated that fossils of *Eldonia*, an ancient organism whose taxonomic affinities are not well understood, are composed entirely of kerogen, most of which is carbon. In a way, this observation correlates with the physical appearance of the Burgess Shale fossils; they are almost always jet-black compressions on the surface of dark gray shale.

So here we have a case where ancient biomolecules, even though they are mostly unrecognizable as individual compounds, actually form the body of the fossil. The Burgess Shale fossils never cease to amaze.

As we have seen in this chapter, ancient biomolecules can shed light on the process of fossilization, as well as help provide accurate classification: chitin was critical to the accurate classification of the sponge *Vauxia,* and an ancient cholesterol-like molecule provided evidence that the Ediacaran *Dickinsonia* was more animal than plant. But as we will soon discover, ancient biomolecules have scientific value beyond their contributions to classification. They can provide us with information about the physiology and behavior of long-extinct organisms—information that would otherwise have been lost forever.

3

THE PURPLE FOSSIL

In Polynesia, a young male palagi* (foreigner) may walk with a single young woman late in the evening but only when accompanied by her brothers and their friends, who follow a very indiscreet 100 feet behind. If the couple stops for too long, stones crash into the brush next to them, a reminder not to get too intimate. The darker the night, the closer the chaperones follow, and the more inaccurate (and persuasive) their throws. Even on a moonless night though, dark is usually not a problem as the sky of the southern hemisphere is unexpectedly bright; it contains many more stars than does the northern hemisphere, including, of course, the Southern Cross. Yet in the pale light, one can see only shades of gray—the dark red hibiscus flower in the young woman's hair is a muted black.

---

* In the Samoan language, "Pa" means split and "lagi" means sky. If you were a Samoan child playing on the beach in 1722, a tiny white speck on the horizon would have presaged the first arrival of a European sailing ship. The speck took on the triangular shape of the ship's sails as it grew nearer and continued to grow in size; it looked like an expanding crack in the sky. Imagine the children running to their village screaming "Palagi! Palagi!"—"The sky is splitting! The sky is splitting!" Hence the Samoan word for a Caucasian or foreigner.

Imagine how bizarre and boring our world would be without its colors—or our ability to see them. Color infiltrates most everything in our lives. For many species, beyond merely making life more interesting, color plays an important role in their very survival: it is used by both animals and plants to attract mates (or pollinators), warn off potential predators, blend into the surrounding scenery, delineate species . . . the list is a long one. But when did pigments—chemicals that are naturally colored—first appear? And which came first, colors or the ability to detect them? One way to answer these questions is to turn to the fossil record. We know, for example, that cyanobacteria, a life-form that goes back billions of years, produced the green pigment chlorophyll. The oldest fossil evidence for color vision, by contrast, consists of the permineralized color-sensitive cone cells of a relatively young 300-million-year-old shark.

However, given that color vision is dependent on the presence of several very specific genes, a better way to estimate the time course of the evolution of color vision may be through the use of a molecular clock. This technique, in its simplest form, assumes that the greater the number of differences (mutations) in the same gene from two different organisms, the less they, and the organisms of which they are a part, are related. Knowing something about the frequency at which such changes occur, one can estimate their ages and the divergence of related organisms over time.

The genes responsible for color vision are called opsins; in most organisms, including humans, three different opsins provide a broad range of color vision. Molecular clock analyses have indicated that all three were present in basal arthropods about 560 million years ago, 50 million years before the Burgess Shale fauna existed.[1] If 560-million-year-old animals had color vision, can we assume that they also produced and displayed colors?

The best evidence to date again comes from the animals of the Burgess Shale. Fossils there, including species of *Wiwaxia* and *Marrella*, had three-dimensional exterior surfaces, which consisted of multiple layers, like a grating; incident light would reflect and diffract from these different layers and produce iridescence.[2] Animals, it appears, have been displaying and seeing colors for over 500 million years, way back in deep time.

Just as colorization of black-and-white motion pictures of the early twentieth century added new dimensions of beauty and reality to those movies, the study of ancient pigments enlivens our understanding of long-extinct organisms. For example, recent study of the ancient pigment melanin has given rise to an exciting new theory about the vision of 500-million-year-old trilobites. Other ancient pigments have been found in organisms as disparate as crinoids, algal reefs, snails, and plants, all of which have fascinating stories to tell.

Visitors to the National Museum of Natural History in Washington, DC, who wander through the Sant Ocean Hall may come upon an exhibit of ancient sea life that beautifully documents the presence of ancient pigments and their function in behavior. The exhibit includes a large piece of Mississippian siltstone on which rests two amazingly well-preserved crinoids. Collected from the world-famous Edwardsville Formation near Crawfordsville, Indiana, they are but two of nearly a hundred different species of crinoids found at that site. Preserved intact, their long stalks and branching arm-like structures (calyces) are sculpted in beautiful three-dimensional detail.

The Crinoidea, today's sea lilies and feather stars, are classified as one of the several groups of echinoderms (a classification that includes starfish and sea urchins) that evolved more than 500 million years ago. Along with trilobites, crinoids dominated the shallow seas of the Paleozoic era for hundreds of millions of years. The crinoids displayed in the museum lived in the waters of a giant delta formed by rivers carrying sediments eroded from the then young Appalachian Mountains about 340 million years ago. The diversity of the Crinoidea was astounding; over 1,200 fossil genera of these marine organisms have been described. Unfortunately, their dominance ended with one of Earth's greatest mass extinctions, the Permian-Triassic event.

The Permian-Triassic extinction occurred quite abruptly over a period of a few tens of thousands of years about 252 million years ago.[3] The oceans became significantly more acidic and much warmer; more than 95 percent of all marine life-foms, including most of the crinoids, disappeared. To understand why this occurred, consider that, when in the field, geologists invariably carry a small vial of acid. When applied to a rock of unknown composition, the production of bubbles of carbon dioxide identifies the rock as limestone. Composed of calcium carbonate, limestone readily dissolves in the acid. The crinoids, mostly stalked sessile organisms that lived in shallow water, had endoskeletons made of calcium carbonate. With the newly acidic nature of the sea water, they were sitting ducks: the water dissolved their endoskeleton and interfered with the production of new carbonate skeletons. This phenomenon is occurring again today as oceans continue to acidify (beyond the 30 percent increase that has already occurred since the onset of the industrial revolution), and it will likely be the downfall of many living marine invertebrates.[4]

Fortunately, a small number of Mesozoic crinoids, believed to have been deep water species, made it through the Permian-Triassic extinction and subsequently diversified into the free-living species that dominate the approximately 600 species that exist today. Many of these living species are brilliantly colored: blue, green, yellow, and red. The intensely colored organisms make no attempt to hide; their color is a warning to predators to stay away—a phenomenon known as aposematism (in Greek *apo* = "away"; *sema* = "sign"). Of course, aposematic organisms must have something with which to back up their bravado. Some biomolecules do double duty, serving as both the pigment responsible for the organism's color, and the toxin that awaits the predator who ignores the crinoid's colorful warning.

There are two different species of crinoids on display in the Ocean Hall, *Elegantocrinus symmetricus* and *Rhodocrinites kirbyi*. While the former is uniformly light gray in color, *Rhodocrinites kirbyi* is dark purple. Nearly 70 years ago, the chemist Max Blumer, then at the University of Basel, described violet-purple pigments from a specimen of the Jurassic crinoid *Liliocrinus*.[5] In the early 1960s, while at the prestigious and beautifully situated Woods Hole Oceanographic Institution, Blumer followed up his initial work with the isolation of a series of pigments from specimens of yet another fossil, the Jurassic crinoid *Apiocrinus*. Forty years later, in the early years of the new millennium, two young students, Klaus Wolkenstein at the University of Heidelberg and Christina O'Malley at Ohio State University, independently picked up Blumer's thread as the topic of their doctoral research.

Wolkenstein, now at the Max Planck Institute for Biophysical Chemistry in Göttingen, Germany, has documented numerous impressive examples of ancient biomolecules through deep time. He has identified ancient pigments in four different orders of fossil crinoids from exposures of Triassic and Jurassic rocks found at different sites around the world. From these fossils, he has isolated at least 10 different fossil pigments, and 20 more that can be found in extant crinoids. One of his most interesting discoveries concerns the presence of a type of purple pigment in two different genera, *Hispidocrinus* and *Hypalocrinus*.[6]

FIGURE 3.1. The Mississippian (340-million-year-old) crinoid *Elegantocrinus sym-metricus,* on display in the Ocean Hall at the National Museum of Natural History, lies adjacent to a second species, the purple *Rhodocrinites kirbyi.*

Although *Hypalocrinus* now lives in the Pacific Ocean along an arc of islands from Japan to New Zealand, *Hispidocrinus* last lived 200 million years ago near what is now the Bristol Channel in southwestern England. What worked in the Jurassic still works today.

The ancient pigments of Crinoidea were amazingly similar to those that exist in the brightly colored echinoderms of the present. Recent pigments, however, contain bromine, an element that is not seen in the fossils. It appears that these Mesozoic fossil pigments, the primary one of which is called hypericin, represent slightly degraded forms of the pigment of living crinoids. Even without bromine, however, the purified ancient pigment is still purple in color. The fact that these pigments have been conserved through the evolution of the modern crinoids over a period of more than 250 million years strongly suggests that they performed essential functions.

If one or more of these hypericin-like pigments were aposematic, what was the mechanism of its toxicity? A clue may come from the flowering plant *Hypericum perforatum*, otherwise known as St. John's Wort, which also makes hypericin. The molecule was first isolated from and named for the plant, the generic name for which was coined by Linnaeus in 1753. It turns out that hypericin has a number of pharmacological effects—it was once thought, for example, to be an antidepressant until clinical trials showed it was ineffective and had potentially dangerous interactions with other commonly prescribed drugs.[7] What effect(s) the hypericin-like biomolecules have on sea urchins, a major predator of crinoids, is unknown. More recent research has demonstrated, however, that several pigment-related biomolecules are toxic to mammalian cells, as they interfere with cell replication.

Compared to the pigments identified by Wolkenstein, the organic pigments reported by Christina O'Malley and colleagues were from several much older specimens, 340-million-year-old crinoids and two noncrinoid echinoderms, the 430-million-year-old *Holocystites* and a specimen of a 450-million-year-old *Agelacrinites*.[8] It's difficult to imagine—original biomolecules, still colored after their isolation from nearly half-billion-year-old fossils. Although O'Malley neither purified the pigments nor clearly characterized their structures, her research indicated that the pigments of closely related echinoderm taxa were more similar to one another than to those of more distantly related groups. Based on these data, she constructed a phylogenetic tree—a figure that shows the evolutionary relationships of different organisms—that supported one of several different proposed evolutionary trees previously established with morphological criteria. While other phylogenetic trees required two independent origins of stemmed echinoderms, her tree supports the idea that there was only one origin. According to the principle of parsimony, which is near sacrosanct in science, we should prefer this simpler explanation. O'Malley's work demonstrates how even ancient biomolecules that do not contain genetic information can help us unlock evolutionary secrets of the past.

The Colors of Ancient Reefs

In the early 1970s, I spent two years as a Peace Corps volunteer on the volcanic island of Savai'i in the South Pacific nation of Samoa. I had been assigned to teach math and science at the Vaipouli School, located a mile up the mountain from the coastal village of Safune. A single dirt road circumnavigated the island, and the only source of electricity was the school's aging diesel generator; the Steinlager beer was cooled by a kerosene-fueled refrigerator. As my teaching duties ended in the early afternoon, I spent a great deal of time snorkeling on the most beautiful reefs in the world. Just a mile northeast of Safune, a small river empties into the ocean; thousands of years of fresh water have killed the coral and cut a channel through the reef that surrounds most of the island. Just east of the channel, however, the reef disappears completely for an 11-mile stretch of Savai'i's coast, buried under a massive lava flow from the 1905–1911 eruption of Mount Matavanu. When the tide is low, and the river continues to flow downhill, the channel provides easy access to the outer face of the reef. Born and raised in northern Minnesota, I knew nothing about the ocean and its currents, and my first foray through the channel was almost my last. Having ridden the channel's surprisingly swift current to the outside of the reef, I belatedly realized that swimming back against that current was impossible. I swam westward, parallel to the crest of the reef; when I looked down, I saw one of the most beautiful reefs in the world.

My exploration of the reef occupied much of my two-year stint on the island. I had never seen a more colorful and diverse array of life. I practiced holding my breath so that I could dive to the bottom of the reef and sit there, as still as possible, and wait for fish and other organisms disturbed by my initial appearance to slowly reassemble and resume their normal behavior. No color was absent. The corals themselves displayed bright reds, yellows, and purples, and the fish, seemingly in competition with the corals, offered up complex multicolored patterns that defined their species. A species of *Tridacna*, the giant clams, displayed the dark royal purple of their mantle. Despite my fascination with the reef and its ability to host such a complex marine ecosystem, I had no appreciation or knowledge of the dominant role that reefs played in the evolution of ancient life.

The reefs that exist today are in large part built by limestone-forming invertebrates (Cnidaria, a phylum that includes the jellyfish and corals) and symbiotic single-celled algae called dinoflagellates. Prior to the collision of the Chicxulub asteroid with the Yucatan Peninsula and the resultant extinction of the majority of Earth's animal and plant species, however, the organisms that built reefs were quite different. Mesozoic and Paleozoic reefs were made by a variety of organisms, including clams, sponges, and species of algae that died off 66 million years ago. The very first reefs were made of stromatolites, horizontal layers or cabbage-like mounds of limestone structures built by bacteria. As discussed in the previous chapter, remnants of these billion-year-old reefs can be easily found in Glacier National Park, if you take the time to look for them. Staring at the huge cliffs by the Hole-in-the-Wall campsite, it is almost impossible to imagine these rocks as colored; could they have had anything like the vivid and diverse rainbow of colors of the reefs I had grown to love in the South Pacific?

From within ancient, 150-million-year-old reefs, scientists have found specimens of the deep time alga *Solenophora jurassica*. Cut a slice through the rock and alternating layers of decidedly pink limestone appear. Commonly called beetroot stone, a polished section of the fossil is a prized possession of collectors around the world. If the rock is powdered and extracted, and if you are in Klaus Wolkenstein's laboratory, a beautiful dark red pigment can be purified from the fossil.[9] Termed borolithochrome (*litho* and *chromo* are Greek for "stone" and "color," respectively), a chemist's rendition of the pigment's structure resembles two small pieces of chicken wire held together by an atom of boron. This is one strange molecule. Boron rarely appears in biomolecules. While similar to the element carbon, it behaves chemically more like silicon. The chicken wire–like portion of the new pigment has been found in only one other organism on Earth, the bacterium *Clostridium beijerinckii*.[10] Wolkenstein speculates that *Solenophora jurassica* may have had a symbiotic relationship with an ancient species of bacterium related to *Clostridium*, though what advantage this relationship provided to either the bacterium or the alga is unknown. What is really bizarre, however, is that the bacterial biomolecule has antibiotic properties. A potential drug, it has now been synthesized in the laboratory. Like the fossil pigment, it is pinkish in color.

Borolithochrome is not, however, the only ancient pigment contained within *Solenophora jurassica*. Holly Barden of the University of Manchester and her colleagues have provided evidence that this fossil also contains a pigment related to phycoerythrin.[11] This pigment, which in extant algae is both red and fluorescent, and functions as a light-harvesting component of the organism's photosynthetic system, is a complex of a protein and a molecule related to chlorophyll. The ancient pigment that Barden's laboratory identified was, unfortunately, no longer intact and was neither red nor fluorescent; it had been degraded. However, the fragment that was preserved shows us that *Solenophora jurassica* harvested the sun's light to fuel its metabolism.

Maryland's State Fossil Snail

Organisms with calcium carbonate support structures make great fossils. These include the stacked plates of a sessile crinoid's stem, the internal skeleton of its feathery arms, and the shells of mollusks. The fossil record would be much poorer if deep time oceans had not been full of brachiopods, clams, and snails. These organisms also provide a huge potential for new discoveries of ancient biomolecules. Snails, in particular, are known to be exceptionally colorful, displaying every color in the rainbow. The brilliant reds, yellows, blues, and greens of more than 50,000 living snail species are sometimes made even more magnificent by their propensity to follow the geometric whorls of the shell to form concentric rings of contrasting colors.

The pigments in many of these snails are called uroporphyrins. Uroporphyrins are a type of porphyrin—which we encountered in the discussion of heme in the abdomen of the blood-engorged mosquito—that were first found in urine (hence, its name). In patients with an inherited congenital disease that interferes with the synthesis of heme, uroporphyrins are produced and color their urine a wine-red color. Porphyrins provide yet another example of nature, as defined herein, having found something that works and subsequently evolving hundreds of variations on the theme. Unfortunately, no one has found ancient uroporphyrin in situ. But scientists are working on it, and beautifully colored snails buried in 10-million-year-old sediments of Chesapeake Bay are a prime target of their investigations.

If you were to drive west on Interstate 68 from Hancock, Maryland, you would soon pass through the massive Sideling Hill road cut. The cut itself is a geologist's delight, as about 800 feet of tilted and multicolored layers of 340- to 360-million-year-old rock are exposed. At an elevation of only about 2,000 feet, it cuts through one of numerous ridges that form the present-day Appalachian Mountains. Formed by tectonic forces, which even now are moving North America towards Japan at the rate of about one inch per year, the Appalachians extend all the way from New Brunswick in Canada to the northern portions of Georgia, a distance of 2,000 miles. The mountains are slowly eroding, sending sediments both east and west. During the Miocene (about 5 to 23 million years ago), these sediments flowed east and settled at the bottom of a huge depression in the earth's crust, a remnant of an 80-mile-diameter crater that formed when a meteor collided with the shallow temperate sea 35 million years ago. This area is now called the Chesapeake Bay.

The famous Calvert Cliffs are essentially a cross section of the Miocene sediments that slid into the bay. As sea levels fell, waves, tides, and storms exposed the sediments at the western edge of the Chesapeake Bay. Along a 24-mile-long portion of the coast of Maryland, this exposure takes the form of a cliff as much as a hundred feet high. This area is also home to the Calvert Marine Museum in Solomons, Maryland. If you ever get a chance, check it out. It is one of the best small museums on the East Coast. John Nance and Stephen Godfrey, the museum's fossil collections manager and curator of paleontology, respectively, are world-class paleontologists who have one of the most famous fossil sites in North America at their back door. Of the many marine organisms that lived in and above the seabed, at last count, over 600 fossil species have been described from the sediments; the site is particularly famous for whales and other marine mammals.[12]

The cliffs at the bay's edge contain numerous recognizable zones or beds that span a period from about 8 to 18 million years ago. Farther west, exposures along the Potomac River are about 23 million years old. Erosion claims as much as three feet of the cliffs each year. Collecting is best after a storm and in the spring when frozen sediments thaw. It is also extremely dangerous; huge blocks of rock regularly fall from the top of the cliffs, as have private cottages with once beautiful views of the bay. It goes without saying that digging into the cliff is a fool's exercise (it is also illegal as the cliffs are mostly located on private property). There are, however, innumerable specimens to be collected on the wide flat beach, including beautifully preserved serrated shark's teeth—just beware of high tides.

The Calvert Cliffs are a very productive source of molluscan specimens with preserved soft tissue. Nance once made a video of a 10-million-year-old quahog (*Mercenaria*) hinge ligament

FIGURE 3.2. Miocene sediments in a cliff south of Calvert Cliffs State Park. Large chunks of rock hundreds of pounds in weight frequently fall from the cliff; these boulders often split from the cliff along relatively thin layers that contain numerous fossil mollusks, including the beautifully spiral *Turritella*.

that stretched like a rubber band, a potential treasure chest of ancient biomolecules. But it was the bright reddish orange snail *Ecphora gardnerae* that first caught his attention. If there was color, might there be a pigment? An ancient pigment? When the Museum hooked up with Robert Hazen of the Carnegie Institute of Science in Washington, DC, they were able to demonstrate that the shell of *Ecphora gardnerae*, once the calcite (calcium carbonate) minerals were removed with acid,

FIGURE 3.3. The Miocene snail *Ecphora gardnerae* collected from Calvert Cliffs, Maryland.

contained relatively large amounts of organic matrix.[13] Their data suggested that the isolated matrix, in the form of very thin sheets of distinctly reddish material, may contain preserved proteins. More importantly, a reddish pigment, yet to be characterized, was present, bound to the organic matrix.

To help characterize the pigment, Nance and Godfrey reached out to Dr. Kliti Grice, who runs a massive interdisciplinary laboratory that has gained worldwide recognition in the field of ancient biomolecules. Grice has often worked with fossil leaves, insects, and other small arthropods, specimens for which sampling was synonymous with destruction. Destructive sampling, as you might expect, is an anathema to most paleontologists, but for Nance and Godfrey, who have collected hundreds of *Ecphora gardnerae* from the Calvert Cliffs, such sampling was seen as an opportunity. Hopefully, the next time I speak to the Calvert Cliffs Fossil Club, John Nance will be able to show me a small vial of reddish orange *Ecphora* shell pigment.

Earth's Most Common Pigment

In a fossil leaf, the preservation of the most minute details of its venation can be startling. Sometimes, there is even evidence of herbivory—the munching of insects wherein, for example, the insect moves along a vein, leaving it in place but eating the adjacent soft tissue to produce a pattern that resembles lace. But nothing could more striking, or surprising, than finding a fossil leaf that has maintained its green color. This rare event recently happened to a lucky volunteer who helped excavate the sediments of an ancient lake in Colorado.

In 2010, the Water and Sanitation District of Snowmass, Colorado, decided to build a reservoir to supply the town with water. The 12-acre site they selected was a swale, a concave area of about 12 acres that, unbeknownst to them, was full of ancient lake sediments. It didn't take long before the bulldozer operator saw huge bones, even tusks, careening off his machine's blade. Ian Miller of the Denver Museum of Nature and Science was called in to assess the situation, and soon after, one of the largest fossil excavations ever done was put in motion. In only nine weeks (on either end of a snowy Colorado winter), about 250 volunteers and scientists from nearly 20 institutions collected over 30,000 fossils; the haul included 52 vertebrate species and about 100 species of plants.[14] Most impressive were the 35 mastodons—males, females, and juveniles. And all of this was obtained from the excavation of just one of the site's 12 acres.

The lake sediments recorded 85,000 years of the lake's existence, from 140,000 to 55,000 years ago. The unearthed mastodons lived in the area from about 115,000 to 100,000 years ago during an interglacial warming event. Mammoths, which were also recovered from the site, were shown to have occupied the area adjacent to the lake from about 87,000 to 60,000 years ago

FIGURE 3.4. The Snowmass excavation site in Colorado in 2010.

during a subsequent interglacial period. It was in these slightly younger sediments that Gussie McCracken, then an intern at the Denver Museum of Nature and Science, recovered leaves from 70,000-year-old layers of peat. The leaves were not vivid green, according to McCracken, but rather a "muted greenish brown that quickly turned dark" as she held them in her hands. Buried in an oxygen-free environment in the absence of light, the leaf had retained its pigmentation.

While it may seem unthinkable that a leaf could maintain its color for tens of thousands of years, fossil leaves that have retained their green color have been reported from much older fossils. Believe it or not, the stomach contents of the extinct tapir-like *Lophiodon tapirotherium*, preserved in 46-million-year-old brown coal near Halle, Germany, contained green, albeit masticated, leaves.[15] Scientists were able to confirm the presence of the ancient pigment chlorophyll in the partially digested leaves. Slightly degraded, it differed little from the original pigment. While the discovery of leaves eaten by *Lophiodon* was a one-off occurrence, millions of leaves, many of them "green," have been preserved in the 16-million-year-old lake sediments exposed near the tiny community of Clarkia, Idaho.

Like the unearthing of the Snowmass fossils, the discovery of the Clarkia fossils was an unlikely though, for paleobotany, very lucky event. Like most fossil sites, it was not discovered by scientists looking for fossils. Entrepreneurial opportunities are often few and far between in northern Idaho, so in 1972, Francis Kienbaum decided to construct a motocross racecourse on his property. Kienbaum used his bulldozer to cut into a small hill as he was creating the track and was amazed to see a piece of newly exposed rock with a beautifully preserved fossil fish. He contacted geologists at the University of Idaho, who quickly realized that Kienbaum had unearthed an important paleobiological window into the past—not so much for fish, but for leaves. Well over a hundred different species of plants have been described from the approximately 16-million-year-old Clarkia site, a number of them new to science.

Not one to miss a business opportunity, Kienbaum's son, Kenneth, established a second business, a commercial fossil site which, for 10 dollars per person, allows the public to dig for fossils. If you are a fan of fossils, you may want to spend some family vacation time at the Fossil Bowl site—though a reservation is highly recommended. The Clarkia fossil site is one of the most easily worked and productive fossil sites that I have ever experienced, and I do not exaggerate when I say that you will be wading up to your knees in fossil leaves. But make sure the trunk of your car is empty when you arrive as you can keep everything you find unless it is something rare or new to the site. (In such cases, Kenneth contacts the University of Idaho.) Visiting paleobotanists who are affiliated with the university can collect on the house.

The rock at the site is soft, and rectangular blocks of oil shale can be split, repeatedly, with an old butter knife, to expose a stunning diversity of leaf types. At the moment of exposure, many of the leaves exhibit a reddish or brownish green color. Unfortunately, as the leaves' surfaces make contact with oxygen, the colors fade, just like the much younger specimens from Snowmass. Many of the leaves are so large and perfectly preserved that one can imagine them as museum specimens. Interestingly, a number of paleobotanists regard the commercial site as an assault, the cause of an irrevocable loss of important scientific information. Every year, the 20-foot-high hillside recedes several feet—how long will this resource last? I don't fault Kienbaum though; in the United States, if you own the land, you own the fossils.

I wanted to examine the fossil leaves at Clarkia for myself, so in the summer of 2019, I made the drive through the Bitterroot Mountain range and into Idaho. More than one study in the scientific literature reported that extracts of Clarkia green leaves produced a green solution, and I hoped to reproduce their results. If you pull a green leaf from a tree, any common alcohol, such as ethanol or isopropyl alcohol (rubbing alcohol), will produce a green solution when used to grind up the leaf. My plan was to try the same procedure with a ground-up fossil.

I contacted Bill Rember, a paleobotanist who has lived near and had studied the Clarkia fossils for over 40 years. Bill did his doctoral thesis on Clarkia plant fossils and was so smitten that, in exchange for geological work for a local garnet mining company, he acquired 10 acres of land along Emerald Creek, which contains a large hill of Clarkia fossils similar in size to that of the

FIGURE 3.5. An orange-colored fossil leaf from freshly split Clarkia shale.

Fossil Bowl.[16] He built a house on the property and has lived among these world-famous fossils ever since. When I arrived, Bill was limping slightly from a fall he had taken the night before as he was chasing cattle away from his quarry. He led me into a pit, about 20 feet deep, that he had carved into the hundred-feet-thick sedimentary rock. The bottom-most two feet of the pit were underwater, but just above that, Bill had cut several pieces of shale, each about a cubic foot in size, with a chainsaw. The rock in which Clarkia leaves are preserved was formed from sediments that settled on the bottom of a lake; the shale is soft, almost like potter's clay.

As Bill split the shale block, leaf after leaf was exposed. He pointed out one that was brown with a slight hint of something you might call sage. "That is about as green as they get," he said. Wait a minute, where was the vivid green? I scraped the leaf into a tiny mortar, added isopropyl alcohol, and ground the leaf into dilute mud. My technique must have been lacking, however, as I came up with nothing. As it turns out, I am not the first person to be disappointed in a search for ancient biomolecules in the Clarkia fossils.

Thankfully, however, my time at Bill's Emerald Creek quarry yielded other rewards. At one point, after splitting a piece of damp shale to reveal a perfectly intact fossil leaf, Bill indicated that the leaf could be recovered intact. This I had to see. We made our way to his house and negotiated our way, via a foot-wide path that meandered through years of accumulated mineral specimens, books, maps, fossils, notebooks, and journals, to the basement, where he picked up a plastic tub of reagents. Once outside and after he had made sure that I was upwind of the proceedings, Bill covered the fossil with an acidic reagent. We waited for several minutes, after which he washed off the reagent and immersed the fossil in water. With a sharp knife blade, Bill proceeded to lift the leaf, intact, from the surface of the shale. I was awestruck. The leaf might as well have been picked up from a damp pile of year-old leaflitter. It was beautifully preserved and as perfectly three-dimensional as any leaf can be. I held in my hand an intact leaf that was 16 million years old.

The sage color of the fossil leaves at Clarkia is due to the photosynthetic pigment chlorophyll, whose function is to capture the sun's energy. Its mechanism is complicated. If it were not, we would be covering our roofs with synthetic chlorophyll pigment instead of solar panels made of silicon. The pigment uses the sun's energy to separate water (H_2O) into oxygen, which is given off as a waste product, and hydrogen ions. The hydrogen ions are then pumped across membranes within packets of pigment called chloroplasts to generate a gradient of electrons, a situation analogous to the way a battery creates a

FIGURE 3.6. Bill Rember lifts a 16-million-year-old leaf from the surface of a Clarkia fossil.

gradient of electrons between its anode and cathode. The gradient forms a reservoir of energy that can be used to, for example, make sugars.

At the pigment's core lurks a ring-like structure similar to the heme we encountered in the abdomen of the fossil blood-engorged mosquito. However, in chlorophyll, an atom of magnesium replaces the iron that is bound in heme, the oxygen-carrying moiety of hemoglobin. A hydrophobic (water-hating) tail of the chlorophyll molecule helps to anchor the pigment in the chloroplast membranes; when cleaved from the rest of the chlorophyll molecule, the tail itself can persist into deep time to produce biomarkers that are of value in the discovery and evaluation of oil deposits.

The green color of chlorophyll is due to its inability to absorb light with wavelengths between about 600 and 500 nanometers; it reflects these wavelengths, and this green light is what we see. Paradoxically, it is within the green wavelengths that solar radiation is at or near its energy maximum; the energy that the pigment absorbs and subsequently uses in the complex process of photosynthesis are the less energetic reds and blues. Why would plants not have developed a pigment that more effectively utilized the sun's most energetic light? Various theories have been advanced to explain this conundrum. One possible explanation is that the sun's radiation at its maximum value is simply too strong and potentially harmful. A second proposal is based on the observation that organisms that were present before the appearance of chlorophyll, such as very deep time archaic microbes, possessed a different photosynthetic pigment that very effectively used, and still uses today, the more energetic green wavelengths. If these microbes were a major component of the oceans' surfaces, green light may not have been available to organisms living among or below these bacteria. The ancestors of plants may have had to use whatever wavelengths they could get.

The fossil record of chlorophyll is quite good. Over 85 years ago, German organic chemists, at that time the best in the world, demonstrated the presence of a chlorophyll degradation product in a variety of coals and oil.[17] Since their reports, oil shales from the 52-million-year-old Green River Formation and coal from the Paleozoic have been shown to contain, as biomarkers, fragments of the pigment.[18] The Paleozoic era, it is worth recalling, ended more than 250 million years ago, and the chlorophyll pigment was still largely intact. And despite my unsuccessful attempt to extract chlorophyll from Clarkia leaves, fossils from a similarly aged site in Oregon that were "vivid green" provided green extracts that contained the nearly intact pigment.[19]

While there are numerous plant pigments—for example, members of the carotenoid and anthocyanin families of pigments are responsible for the colors of carrots and blueberries—these pigments have a very poor fossil record. They are very susceptible to degradation mediated by a variety of agents (e.g., metals, heat, oxygen, ultraviolet light, and acids). After just 24 hours in daylight, 50 percent of a given amount of beta-carotene is destroyed. Even when kept frozen and in the dark, only about 1 percent survives intact after a month's time. However, evidence for a pigment related to carotene has been found, not in plant fossils but in the fossil of a 100-million-year-old snake from northeastern Spain described by Maria McNamara and her colleagues.[20] McNamara, an expert on the preservation of soft tissue and ancient color, identified the ancient pigment not through chemical analysis, but rather by the presence of fossilized pigment-specific storage structures in the fossilized skin of the snake. Histological examination of the skin revealed structures that closely resembled those in a recent snake. Unfortunately, the actual identification and chemical characterization of ancient pigment molecules in fossil vertebrates has yet to be reported. That is, except, for one important exception: melanin.

How much more interesting life would be if humans displayed the array of colors that many other organisms do. Unfortunately, we produce only one, melanin, although we can make various related forms of that pigment. These melanins are responsible, in part, for the color of our eyes and the black and red colors of our hair. In humans, that's it, our entire repertoire of pigmentation. Fortunately for us, the melanins, in both humans and a wide variety of other organisms, have an impressive fossil record. With the exception of ancient DNA, melanin pigments may represent the most exciting area of ancient biomolecule research. Pigments are being chased into deep time by numerous prominent and very active laboratories, their findings frequently reported in newspapers and blogs around the world. The reason for this interest, as I will soon discuss, is that pigmentation is a critical component of behavior (e.g., sexual displays) and has functioned that way for hundreds of millions of years.

4

THE BLACK PIGMENT

The detection of an ancient biomolecule that has survived in situ for millions of years is, in and of itself, a remarkable achievement. The ideal, however, is to use that ancient biomolecule to learn about the phylogenetic, behavioral, or physiological aspects of the organism of which it was once a part. Few small molecules have more promise of doing just that than pigments. As we have seen in the previous chapter, nature has produced a wide variety of pigments; however, humans and most other mammals tend to have just one: melanin. Melanins are responsible for the color of our skin, whether innate or suntan-induced, and function to absorb mutation-causing ultraviolet radiation.[1] Contact with melanin dissipates nearly 90 percent of ultraviolet radiation's energy, as heat, within a billionth of a second. In addition to their role in skin color, melanins are also partly responsible for the color of our eyes and hair. The fossil record of melanin, however, is primarily found in organisms much older than the extremely young genus *Homo*.

Scientists around the world are, for instance, working diligently to document the feathered dinosaurs' repertoire of melanin-based pigments, the colors those pigments produced, and the color-dependent behaviors of Dinosauria. We will encounter these exciting discoveries in the following chapter. However, to appreciate the role of pigmentation in the lives of dinosaurs and other organisms, it will help to first explore equally important and fascinating subjects: the evolution of melanin itself, its distribution throughout the animal kingdom (it appears in nearly all animals, from insects to vertebrates), and its function in ancient physiological processes, such as thermoregulation and vision.

Melanin is thought to have first appeared not as a pigment but as an antioxidant. Since then, it has evolved numerous additional functions. It inhibits the growth of bacteria, it is a structural element (it contributes to the hardness of feathers), it protects against ultraviolet light, and it is essential to vision. As a pigment, it functions in sexual display, thermoregulation, and crypsis (camouflage); in addition, when an animal can't hide, it offers colorful warning signals to potential predators. In insects, melanin is not only a pigment. It is also a major player in what passes for an immune system. Melanin is produced at wound sites to produce a Band-Aid of sorts, analogous to a vertebrate's blood clot. In a process called melanization, the pigment is used to encapsulate viruses, microbes, and parasites. For example, mosquitoes are known to inter malarial parasites in tombs of melanin. Scientists are convinced that melanin has additional unknown functions, as evidenced by its presence in internal organs such as the spleen, liver, and kidney. An indication of its ubiquity and importance to life on Earth is the existence of over 600 genes that are involved in melanin synthesis and function and even the pigment's transport and storage.[2]

When arthropods such as crabs and insects molt, they discard their old exoskeleton and replace it with a new one. As any fan of Chesapeake Bay soft-shell crabs knows, the new exoskeleton is soft and succulent—and $150.00/dozen at the fish market on the Potomac River in Washington, DC. Only later, through the process of sclerotization (derived from the Greek *skleros*, meaning "hard"), in which components of the exoskeleton are extensively cross-linked, does it harden. The chemical reactions responsible for sclerotization are thought to have originated over 750 million years ago, before the Cambrian explosion, before the appearance of the Ediacara fauna.[3] Perhaps way before, in ancient fungi. It's not much of a stretch to think that sclerotization was a major enabling factor in the subsequent radiation that made arthropods the most diverse of Earth's complex life-forms.

Important to our story is that, in many insects, the exoskeleton often darkens as it hardens, sometimes to a pitch-black color, like that of a house cricket. The processes of sclerotization and pigment production are integrally related. This black pigment is melanin, a biomolecule as fascinating as its fossil record is controversial.

The pigment melanin has been identified in dozens of extinct organisms, including fossil birds, frogs, bats, and fish ranging in age from 3 to 300 million years old.[4] It was first chemically identified in the ink sac of a 195-million-year-old cephalopod (the class containing octopuses, squid, and cuttlefish).[5] The compressible ink sac organ enables the organism to produce a defensive cloud of dark ink, which acts as a smoke screen and allows it to escape predators. Obviously, the darker the ink the better. Detection of melanin-based ink in fossil cephalopods dates this unique defensive behavior to well over 200 million years in the past and is a wonderful example of how ancient biomolecules can help us understand the evolution of complex behavioral and physiological processes.

Ancient Melanin and Behavior

A very small fraction of fossils depict behavior. Mating flies, preserved both in amber and as compressions, are favorites of fossil collectors. A beautiful specimen from the Eocene Messel site in Germany portrays two 50-million-year-old turtles, *Allaeochelys crassesculpta*, in copula.[6] Predation is preserved in numerous fossils from the Eocene Green River Formation, which depict fish that choked to death while trying to swallow other fish. In a famous 80-million-year-old specimen from Mongolia, the herbivore *Protoceratops* is locked in combat with a *Velociraptor*, the carnivore's scythe-like claw slicing through the throat of the *Protoceratops*.[7]

Although direct documentation of ancient behavior is rare, we can predict such behavior based on our knowledge of ancient color patterns. Color provides evidence for a host of behavioral adaptations, including the phenomena of crypsis (detection avoidance), deimatism (threat behavior), and aposematism (signals of warning). Consider, for instance, the 193-million-year-old ichthyosaur *Stenopterygius quadriscissus*, which appears to have been uniformly dark, if not black, in color.[8] Perhaps, as it dove deep into the sea, the ichthyosaur was less visible (and, therefore, a more effective predator) as a result of its pigmentation. Beyond identifying color-based behavior in a single species, ancient melanin can also shed light on the broader trajectory of life on Earth: if paleobiologists find large numbers of fossils of closely related species that existed over a relatively short period of time, they can follow the evolution of color-based behaviors.

One recent discovery regarding ancient behavior involved a particularly fascinating fossil preserved in the shale rock of a site near the ancient city of Teruel in northeastern Spain. It is about 10 million years old, but, given that it was collected as a curiosity by miners, the exact layer of shale in which the specimen was embedded is unknown, and the date therefore a little fuzzy (10 ± 1.5 million years). The specimen of interest was the fossilized skin of a snake—no bones, just the skin. Because the original tissue had been permineralized, the organic components replaced by calcium phosphate, the skin's ancient biomolecules, including its melanin pigments, were absent. However, minerals had replicated even the smallest anatomical features with incredible fidelity. At the microscopic scale, epidermal and dermal epithelial cells and even collagen fibrils had been duplicated in stone. Also preserved were the storage vesicles in which the skin's pigments had been stored.

Melanin and other pigments are compartmentalized and stored in microscopic three-dimensional packets. For example, melanophores (or melanosomes) store melanin, and xanthophores store carotenoid pigments. Maria McNamara, one of the foremost researchers of a young cohort that constitutes the leading edge of ancient biomolecule research, has helped to establish that these microscopic packets are preserved in the fossil record. Working at University College Cork, she and her colleagues have shown that the numbers, sizes, shapes, and positioning of the packets of melanin and other pigments within the dermis of extant snakes can predict external color patterns. Given the beautiful preservation of the packets of pigments in the fossilized skin of the ancient Spanish snake, McNamara's team was able to characterize them and propose a color pattern for the snake.[9] They eliminated red, blue, and dark green colors and suggested that the back of the snake was decorated with a series of about 30 diamond-shaped patches, black/brown in color on a light greenish background. The pattern of pigmentation

FIGURE 4.1. An example of convergence. A forewing of *Bicyclus anynana*, the extant brush brown butterfly on the left, and, at the right, the 125-million-year-old fossil *Oregramma illecebrosa*. The latter insect belongs to the order Neuroptera, which is more closely related to beetles than butterflies.

indicates two important behavioral facts about the snake. It was active during the day—you don't need camouflage if you're nocturnal—and it could avoid predators by a combination of motionlessness and camouflage.

Melanin has also helped us understand the behavior of ancient insects. Conrad Labandeira, curator of fossil arthropods at the Smithsonian, has recently identified melanin in the multicolored eyespots of the wings of an array of extinct insects from northeast China (*Oregramma illecebrosa*).[10] Related to today's lacewings, they are easily the most impressive fossil insects I have ever seen. Several inches across and beautifully preserved, these fossils would appear to most people to be fossils of butterflies. Look more closely and you would see, in addition to the eye spots, a long coiled proboscis and wing scales, both characteristics of butterflies and moths. But these insects are completely unrelated to butterflies. Ranging in age from 125 to 165 million years old, they lived tens of millions of years before butterflies first appeared on Earth. The existence of eyespots on these Jurassic insects, sometimes called the "butterflies of the

Jurassic," as well as on today's buckeye butterfly, is a beautiful example of convergence—the independent evolution of similar traits in unrelated species (another example: the prehensile tails of New World monkeys and marsupial opossums). In both cases, the eyespots are thought to function as a bluff; to mimic the eyes of a much larger organism and scare off potential predators. What kind of predators was *Oregramma illecebrosa* trying to frighten away? Our answer to that question will have to await another discovery by an enterprising paleobiologist.

Ancient Melanin and Physiology

The origin(s) of vision has always fascinated evolutionary biologists. We know that it first appeared way back in deep time. The external surfaces of *Wiwaxia* are thought to have produced colors by a rainbow-like scattering of incoming light 515 million years ago, which suggests the presence of not just vision, but color vision.[11] Molecular biologists, using the molecular clock, have calculated that the genes responsible for color vision evolved about 560 million years ago.[12] Recent work suggests, however, that different forms of color vision may have evolved on more than one occasion. Dipon Ghosh and his colleagues at Yale University have demonstrated that the roundworm (Nematoda) *Caenorhabditis elegans* actively avoids bacteria that produce a bright blue pigment.[13] Experiments have shown that they will even flee from *E. coli*, bacteria on which they normally feed, when it has been genetically engineered to make the pigment. Yet *Caenorhabditis elegans* does not possess any genes that are even remotely related to the opsins that form the basis of color vision in other organisms. Other primitive organisms also show color-responsive behaviors in the absence of opsin genes—evolutionary remnants, perhaps, of deep time experiments with color vision.

The more familiar organ of vision, the eye, has a fossil record that spans more than a half billion years, as exhibited by many fossils of the Burgess Shale fauna. There is even a fossil record of color vision, in the form of fossilized cone cells in the eye of the 300-million-year-old fish *Acanthodes bridgei*.[14] Melanin has played a role in vision for more than 309 million years, although its precise role depends on the structure of the eye. In vertebrates, the pigment's role is to absorb potentially damaging ultraviolet light, while in insects recent research indicates that it functions to seal off individual compartments (ommatidia) of the eye from light that enters adjacent compartments. The deep time association between the eye and melanin has made for some very productive paleobiology. Recently, this association has led scientists to ask new questions about a very ancient organism indeed, the trilobite.

Discovery of the Burgess Shale fossils would have been delayed for many decades if it had not been for the trilobites of Mount Stephens. Across the Kicking Horse River from the Burgess Shale quarry, the Mount Stephen fossil site is nearly 700 feet lower than the more famous locale. Nevertheless, reaching the slopes of Mount Stephen requires a three-hour, five-mile hike. When you arrive, you will find that the thin layers of shale are covered by 15 different species of trilobites. Examine the fossils carefully, and you will see eyes. These trilobites evolved complex compound eyes, which consisted of numerous hexagonal cells, called ommatidia. Each cell was covered with a hexagonal lens that has always been thought to be composed of a single transparent crystal of calcite. This belief was based partly on the fact that there is a modern species of brittle star that has very small calcite lenses, and partly on the fact that no one had found melanin in a trilobite eye.

The compound eyes of insects, some of which contain hundreds of hexagonal lenses, are impressive structures. I remember the first time I looked at the eye of a Green River Formation crane fly under the magnification of a microscope; the fidelity of the preservation was a revelation. Johan Lindgren of Lund University in Sweden also looked at the eyes of a fossil crane fly, but he saw what I did not. The fossil fly, collected from the 54-million-year-old Fur Formation in Denmark, contained melanin.[15] When he and his colleagues examined the fossil with a scanning electron microscope, they found a thick layer of the pigment surrounding each hexagonal lens. The layer of melanin extended down and around the exterior of each ommatidium to cover the six sides of each hexagonal cell. Since each ommatidium acts as a separate light-sensing structure, if the light that entered one lens scattered into adjacent ommatidia, it would interfere with the organism's ability to create a focused image. Melanin, which efficiently absorbs a wide range of wavelengths of light, effectively prevents light that enters one ommatidium from entering an adjacent one.

Lindgren and his colleagues saw even more. The fossil ommatidia contained hexagonal lenses made of single calcite crystals. In this case, however, they were able to show that the crystals were the product of mineralization that occurred as part of the fossilization process. To prove this, they examined living species of crane flies and demonstrated that the lenses of their eyes were made of chitin. The calcite crystals in the fossil were an artifact of preservation. They then asked the obvious question: Are the calcite crystals of trilobite eyes also an artifact? While Lindgren believes that they are, his interpretation remains highly controversial.[16] Unfortunately, there are no living trilobites to examine.

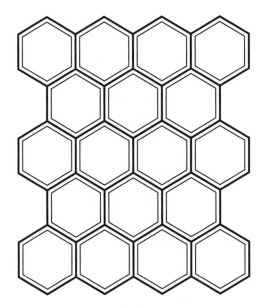

FIGURE 4.2. A diagram of a portion of a fossil crane fly eye showing the surfaces of some of the many hexagonal ommatidia, the lenses of the compound insect eye. Each lens is surrounded by a jacket of black melanin pigment.

While melanin has been a vital component of the eye in many species, in the 54-million-year-old juvenile sea turtle *Tasbacka danica,* it appears to have played an entirely different physiological function.[17] Discovered in a limestone concretion on the Isle of Mors in Denmark, the fossil is spectacular: it looks like a turtle that you would find in a pet store. It is tiny, only about three inches in length, a hatchling. Based on its age and darkly pigmented dorsal shell, Lindgren and his colleagues have concluded that the pigmentation functioned to absorb sunlight and help regulate the young turtle's temperature—just as it does in living turtles. The hatchlings of most living species of sea turtles are known to rest on the water's surface with their back exposed to the sun. Laboratory experiments have shown that the sunlight absorbed by the darkly colored surface of the turtle's back raises its body temperature. Tennis whites are white for a reason. As the hatchlings' temperature increases, its metabolic rate increases, and as a result, it grows faster.[18] As sea turtles mature, they lose much of their dark pigmentation, as their ancestors have done for the last 54 million years.

Ancient Melanin and Phylogeny

Mazon Creek nodules, and the fossils inside them, average several inches in length. One of the most famous exceptions is *Tullimonstrum gregarium*, specimens of which can be over two feet in length. Slender, with a broad tail that resembles the vertical tail wings of a World War I fighter plane and a long anterior appendage sporting a pincher-like organ, this organism was weird. It soon acquired the name Tully Monster. Paleobiologists still don't fully know what it is. At different times, it has been described as a worm, a mollusk, an arthropod, and a vertebrate. Its pair of stalked eyes, which sat on top of its body just behind its head, recently became the focus of a study by a team of scientists from the Universities of Leicester, Bristol, and Texas at Austin.[19] A well-preserved specimen had eyes that contained melanin and melanosomes arranged in layers like those found in the pigmented epithelium of the vertebrate retina. Given that pigmented retinal epithelium is found only in vertebrates, the researchers concluded that the Tully Monster was also a vertebrate. Criticism soon followed. Another team of scientists argued that such an assignment must be based on numerous vertebrate-specific characteristics.[20] They also argued that many invertebrates have a pigmented epithelium that lines the surface of an involuted cup-like structure, a kind of intermediate stage in the evolution of the more complex eye with which we are most familiar.

The *Tullimonstrum gregarium* controversy—is it a vertebrate or an invertebrate?—was eventually settled with additional studies of ancient biomolecules. When the soft tissues of organisms are preserved as compression fossils, they sometimes consist of a very thin layer of organic matter rich in carbon—they are said to be carbonaceous. Chitin, for instance, is rich in carbon. In fossils of invertebrates that have a chitinous exoskeleton, much of the fossil is sometimes composed of derivatives of the polymeric sugars that make up chitin. However, in vertebrates, major structural elements are made of protein (e.g., bone collagen), and derivatives of that protein are also sometimes preserved in compression fossils. There is, however, an important difference between the derivatives of chitin and proteins: degradation products of proteins are rich in sulfur while the degradation products of polymeric sugars (polysaccharides) are, for the most part, sulfur-free. In theory, one should be able to discriminate between the degradation products of chitin and proteins and use this finding to differentiate invertebrates from vertebrates.

Tory McCoy and Jasmina Wiemann, students in Derek Briggs's laboratory, decided to see if a technique called Raman spectroscopy* could discriminate between chitin- and protein-derived ancient biomolecules. They applied this technique to deep time fossils that had been unequivocally identified as either invertebrate or vertebrate—including arthropods with

---

* The technique is named after its inventor, Chandrasekhara Raman, who first described the physics of the light scattering that underlies the technique, work for which he won a Nobel Prize.

exoskeletons and the egg capsule of a cartilaginous fish, each from Mazon Creek. The process wasn't straightforward, as the ancient derivatives were numerous and varied. In the end, however, they were able to establish that invertebrate and vertebrate soft tissues could indeed be reliably differentiated. They then applied the technique to *Tullimonstrum gregarium*. The data demonstrated that the Tully Monster was a vertebrate.[21] Morphological analyses of nearly 1,200 specimens of *Tullimonstrum gregarium* in Briggs's laboratory led to the conclusion that this ancient organism was an ancestor of lampreys.

In 2016, melanin was identified in a 307-million-year-old specimen of the hagfish *Mayomyzon pieckoensis,* which was also collected from the iconic Mazon Creek sites in Illinois. Modern hagfish, along with lampreys, make up a group of animals called cyclostomes (Greek for "round mouth"), which lack jaws. The two groups differ, however, with respect to their eyes: lampreys have a relatively advanced and complex eye, while hagfish lack such structures as a lens, iris, and retina. As a result, it has been suggested that the hagfish eye is a primitive intermediate in the evolution of the more complex visual organ characteristic of most other vertebrates. However, when the eye of *Mayomyzon pieckoensis* was examined with scanning electron microscopy, it was shown to contain both melanosomes and melanin. The eyes of the fossil hagfish were more modern than had been thought. Molecular analyses have demonstrated that lampreys and hagfish are closely related, but the hagfish eye, for reasons unknown, degenerated through hundreds of millions of years to a more rudimentary form.[22]

This work demonstrates, yet again, that ancient biomolecules that do not contain genetic information can still be of immense scientific value, even in the fields of phylogeny and taxonomy.

The examples that we have discussed are a drop in the bucket compared to what awaits us in the near future. There is a new science in town, and what better way to experience it than through the courtship displays, thermoregulatory strategies, predator-prey interactions, and evolution of everyone's favorite extinct organisms, the feathered theropod dinosaurs.

5

DINO FEATHERS

Recent studies have provided unequivocal evidence that many dinosaurs, even those that were not direct ancestors of the therapods, had feathers—in many cases, brightly colored feathers. Our ability to describe the pigmentation of those feathers presents an opportunity for the world's adventure parks, filmmakers, and museums to revamp their exhibits and portrayals of dinosaurs, often some of their most popular attractions. For instance, in 2019 alone, nearly three million people visited the National Museum of Natural History in Washington, DC, many to see the new Deep Time exhibit in the David H. Koch Hall of Fossils. More importantly, for our purposes, our ability to describe this pigmentation provides the potential to better understand dinosaur behavior and physiology.

The colors of dinosaur feathers are produced by two different phenomena: those based on pigments, and those that are based on the microarchitecture of the feather's surface. The latter involves the scattering of light from multiple micro-thin layers of biomolecules at the skin's surface. Much like a rainbow, which is produced by the scattering of light as droplets of water hanging in the air diffract the sun's light, the iridescent colors that result are spectacular. Ancient fossil gratings are almost always organic in nature, and they can be replicated with such precision that they still function to scatter light. The other source of color production in dinosaurs derives from pigments, specifically melanins.

About a decade ago, research on ancient melanin underwent an explosive acceleration as the result of two unrelated discoveries. The first was the discovery of what are arguably the richest fossil beds in the world, especially as they relate to the evolution of birds. While we have known for more than a quarter of a century that some dinosaurs had feathers, our understanding of the variety and interrelationships of feathered dinosaurs and early birds has increased dramatically due to fossil sites in northeastern China. Situated along the Yellow Sea between North Korea and Beijing, Liaoning Province is home to a number of fossiliferous deposits of different ages. For example, the Yanliao fossils date to about 166 million years ago, while the Jehol fossils date from 120 to 133 million years ago.[1]

The fossils from China are exceptional in several ways. They are frequently completely articulated skeletons as opposed to individual randomly distributed bones. Moreover, the shale in which they are preserved often splits to reveal not a single fossil but what paleontologists refer to as "part and counterpart"— two views of the same organism. As the rock is split, the fossil itself splits in half to produce fossils on both faces of the newly exposed rock. Having two versions of the same fossil is better than one, as each half often displays features absent in the other. Perhaps most importantly, "soft tissues," such as feathers, are preserved at several fossil sites in China. Not only is the quality of the fossils in these sites exceptional, the raw number of fossils is simply astounding: over a hundred new species of dinosaurs and birds have been found, and most species are represented by numerous specimens.

The second discovery was made by Jakob Vinther while he was a Ph.D. student in the laboratory of Derek Briggs at Yale University. In 2008, he realized that what scientists had thought for decades to be fossil bacteria were actually fossils of tiny packets of melanin.[2] Called melanosomes, these small structures, which vary in both size and shape, had been known to exist in vertebrate tissue for nearly a century. It wasn't until very recently, however, when melanosomes were shown to be present in spiders, that scientists understood how broadly distributed melanosomes were across the animal kingdom.

Although many details of the process of melanosome formation remain unclear, it appears that storage of melanin within melanosomes is dependent on the presence of amyloid proteins. White Leghorn chickens, for instance, are pure white as a result of mutations in a melanosome amyloid protein. And this effect can be reproduced in the laboratory: engineered genetic mutations can transform a dark-haired mouse into one without any pigmentation whatsoever. The term "amyloid" does not refer to one particular protein but rather any protein that, often as the result of mutations, can alter its shape. As its shape changes, it becomes "sticky" and more readily adheres to other molecules of the same protein. In the case of melanosomes, these proteins form very ordered arrangements, but in other cases they self-assemble into masses of disorganized fibers. Amyloid proteins are, for example, one of the culprits responsible for the brain lesions seen in Alzheimer's and Parkinson's disease patients. The plaques found in patient's brains are simply deposits of thousands of aggregated and therefore insoluble amyloid proteins.

Many of the fossil feathers that Vinther, Briggs, and their colleagues examined consisted entirely of ancient melanosomes. The feathers were literally a mass of melanosomes, which they immediately recognized could shed light on the color patterns of feathered dinosaurs. Their subsequent work on dinosaurs from China and the iconic 150-million-year-old *Archaeopteryx lithographica* from Germany began with the observation that melanosomes come in different shapes: rod-shaped melanosomes that contained black melanins and spherical melanosomes full of melanins responsible for yellow to reddish colors.[3] Despite the simplicity of this observation, however, the process for determining the color of a feather is exceedingly complicated.

The insoluble melanin molecule is so big and so complex in its makeup that it is impossible to isolate as a single biomolecule. As a result, scientists do not understand the exact structure of polymeric melanin, nor can they identify intact melanin as a single peak on a graph issuing from an analytical instrument. To make matters worse, there are different types of melanin with distinct subunits that are incorporated into melanin polymers. Those that contain sulfur combine to form melanins that are responsible for yellow to orange to red colors. Melanins without sulfur produce black and shades of gray. But it is not an either-or affair. Orange feathers of the extant zebra finch have 25 percent sulfur-containing subunits and 75 percent sulfur-free subunits.[4] Even pitch-black hair contains as much as 15 percent sulfur-containing subunits. In fact, the two different types of melanins are nearly always found together. And to make it even more difficult for the paleobiologist, there is more than one type of each of these two pigments, and different colors can be produced by different ratios of the various kinds of melanin pigments!

One preferred analytical technique is to vaporize and fragment the constituents of a fossil feather, with the hope of finding small pieces of the complex polymeric pigment that will provide a diagnostic fingerprint-like signature for each unique melanin-based color. To work, the analyses must be quantitative: we need to know not merely which subtypes are present, but also the exact percentages of each melanin subtype. To make things even more difficult, the analysis must be nondestructive. No one is going to destroy an irreplaceable specimen in order to find out what color it was. Very recently, Ismael Galván, Kazumasa Wakamatsu, and their colleagues developed a noninvasive method to measure the ratios of some of the different subunits in the melanins of feathers of living birds.[5] They found that the ratio of the sulfur-free subunits in black feathers differed by only about 10 percent from those in gray feathers. Given this subtle difference, application of their work to the ancient and often degraded melanins of deep time fossils is going to be difficult.

Nature's rationale for the production of different sizes and shapes of melanosomes remains a mystery. The correlation between color and melanosome shape and size does not appear to hold in the skin of lizards, crocodiles, or turtles, where melanosomes show very little variation in shape and size and provide no information about color. Some scientists argue, moreover, that a tight correlation between melanosome size and shape and feather color may present too tidy a picture. For one thing, there is a great deal of overlap among the sizes and shapes of melanosomes in extant feathers of a single color. In addition, the melanosomes of some feathers have been found to be hollow, but they only become so as the feather grows. In other words, melanosome morphology can change over time.

Things become even more complex when we examine fossils. Measurements of fossil melanosomes are often made not on preserved melanosomes but on the impressions or molds that were left in the rock matrix after the melanosomes degraded

FIGURE 5.1. Melanosomes come in many different sizes and shapes, some even hollow, and contain highly variable mixtures of melanins.

and disappeared. The size and shape of these impressions may be subject to shrinkage and distortion during the fossilization process. Fully aware of these potential complications, paleobiologists have nevertheless pressed on. Scientists in several laboratories have measured the shapes and dimensions of melanosomes of living birds, which they have attempted to correlate with the known colors of their feathers, an example of paleobiology driving neontological research (the study of living organisms— *neo* and *paleo* from the Greek words for "new" and "ancient," respectively). In one study, scientists determined the shapes of melanosomes in feathers of various types and colors from 232 different living birds; melanosome shapes and sizes varied from small and circular to large and elongated.[6] In a second study, melanosomes of iridescent feathers in 97 different species of extant birds were examined.[7] In both cases, statistical analyses of the size and shapes of the melanosomes suggested that these parameters could, if correctly applied, be used to assign colors.

The Colors of Dinosaurs

What better place to begin an account of ancient dinosaurian color than with the 150-million-year-old *Archaeopteryx*, one of the most significant fossils ever found. Most scientists agree that, while not a direct ancestor of birds, it was likely a distantly related cousin. *Archaeopteryx* had feathers, but it is unclear if it was capable of flight, as it lacked the large, keeled sternum (breastbone) that provides an attachment site for large flight muscles. We do know, however, that the feathers of *Archaeopteryx* were colored. Two separate studies by Matthew Shawkey and his colleagues, based on analyses of a rare specimen that contained both intact fossil melanosomes and their imprints, have demonstrated with greater than 90 percent probability that *Archaeopteryx* was black.[8]

Paleobiologists have shown that the sizes and shapes of melanosomes increased significantly during the transition from cold-blooded (ectothermic) to warm-blooded (endothermic) animals.[9] Was there a causal relationship between these two changes? While there are theories that link melanosomes with metabolic rates, there is no direct fossil evidence to support these theories. There is, however, data that support a role for melanization in flight. Increased levels of melanosomes have been shown to correlate with feather strength. And in a study with living birds, Matthew Shawkey and his collaborators have shown that the black wing tips of white birds are warmer than the rest of the wing, which, as a result, produce increased lift and flight efficiency.[10] It is not at all fanciful to suggest that melanization played a role in the evolution of flight in birds.

Studies of pigmentation in other dinosaurs, both those unrelated to therapods and those considered true birds, often reveal that these creatures were even more spectacular than we previously imagined. Consider, for instance, *Anchiornis hux-leyi*, which was recovered from the Daxishan site in Liaoning Province. Completely covered with feathers, the foot-long, 155-million-year-old specimen contained preserved melano-somes with a variety of three-dimensional shapes. Analysis indicates that the body of the bird was gray, the upper surface of its wings was made of parallel stripes of dark black and white and, most impressive, a long crest of feathers on the top of its head was bright red. The researchers responsible for this finding have suggested that the pigmentation of the wing may have served as a sort of camouflage, while the bird's beautiful red crest feathers must have been a signal in sexual display/court-ship behavior.[11]

Equally impressive is *Caihong juji*, a 161-million-year-old hollow-boned and feathered bird-like theropod, again from China, which had a bony crest on its skull covered with feathers. When the melanosomes of the crest's feathers were examined in a scanning electron microscope, they were shown to be flattened (pancake-shaped) and arranged in parallel layers.[12] Unlike the shapes of melanosomes that are known to be responsible for the usual array of colors, the function of these uniquely shaped melanosomes was a mystery—that is, until it was discovered that the brilliant iridescence of feathers of extant humming-birds is produced by similarly shaped melanosomes. The irides-cent feathers on the head and crest of *Caihong juji* document the earliest known iridescent dinosaur.

But even dinosaurs without feathers appear to have been col-ored. Given what we know about skin pigmentation's role in thermoregulation and behavior in extant organisms, we must assume that skin color has been around for hundreds of mil-lions of years. Take, for example, the huge plant eater *Borealo-pelta markmitchelli*, which is more closely related to *Triceratops* than predatory therapods (although the two herbivores were separated by tens of millions of years). Discovered in Alberta, Canada, the epidermis of a 120-million-year-old fossil of *B. markmitchelli* was preserved as a thin layer of organic material.[13] Unfortunately, melanosomes had not been preserved. However, Caleb Brown, working at the Royal Tyrrell Museum in Drumheller, Alberta, was able to identify a small fragment of the melanin pigment itself that proved to be diagnostic of a type of melanin, called pheomelanin, responsible for yellow, orange, and red colors. The analyses by Brown and his colleagues suggested that the dinosaur's body displayed two parallel horizontal zones with a lighter red color below a brown/red back. Its huge, armored neck sprouted multiple rows of large black spines.

Such colors are an example of countershading: color patterns that help an animal to disappear into the background. But *Borealopelta markmitchelli* was 15 feet long and weighed about 2,500 pounds. From what predators did they have to hide? The largest land mammals that live today, such as the elephants, have no need for countershading—their size alone is sufficient defense. The Cretaceous period, however, was full of arguably the largest and most fearsome predators ever to have existed. Although the predatory theropod *Acrocanthosaurus*, which inhabited North America at the same time as *Borealopelta markmitchelli,* was slightly smaller than *Tyranosaurus rex*, it was still six times the size of the pigmented herbivore. *Borealopelta markmitchelli* had armored plates and long sharp spikes but probably preferred to simply avoid trouble entirely. The tactic of disappearing into the background also appears to have been adapted by dinosaurs to protect their eggs. This recent discovery was made possible by a famous find that took place over 40 years ago.

Jack Horner was a preparator at Princeton University, but every year he and his good friend Bob Makela, a high school teacher in Rudyard, Montana, would spend their summers doing fieldwork. In 1978 Horner became aware of the existence of a specimen of what was obviously a dinosaur hatchling when he visited the Trex Agate Shop in the crossroads hamlet of Bynum, Montana. Horner and Makela subsequently published a description of *Maiasaura peeblesorum*, a 77-million-year-old duck-billed dinosaur and its nest, a meter-wide crater-shaped mound of dirt that contained dozens of eggs and hatchlings.[14] It was a revelational discovery, in that it clearly demonstrated parental care—the eggs were not simply laid and abandoned with the assumption that the sun and luck would see the young through. One question, which remained unanswered, was whether the dinosaur eggs were colored.

Of all the vertebrates that lay eggs, birds are the only ones that produce colored eggs. It is widely held that the pigmentation, which includes patterns or speckling, helps to camouflage the eggs and protect the nest from predation. The pigments in the eggs of living birds have been well characterized. There are essentially two: protoporphyrin, the structure of which is similar to that of the porphyrin found in hemoglobin, and biliverdin. In humans and other vertebrates, biliverdin is a breakdown product of the heme in senescent red blood cells, and as a result, it is structurally related to porphyrin. Despite their structural similarity, different combinations of these pigments account for the diversity of bird egg pigmentation. Given that birds are directly descended from avian dinosaurs, did these dinosaurs also have pigmented eggs? If so, when did egg pigmentation evolve? Forty years after Horner's discovery, Jasmina Wiemann and her colleagues were able to provide an answer.

Wiemann has been fascinated by molecular paleobiology since her undergraduate studies at the University of Bonn. Working with the paleobiologist P. Martin Sander, both as an undergraduate and as a graduate student, she became familiar with cutting-edge analytical techniques, including Raman spectroscopy. Wiemann's interests and expertise attracted the attention of Derek Briggs. As a graduate student in Briggs's laboratory, Wiemann's expertise has resulted in several important breakthroughs, including, as previously discussed, the demonstration that the Tully Monster was a vertebrate.

Wiemann and her colleagues decided to use Raman spectroscopy to examine avian and nonavian dinosaur eggshells in an attempt to detect ancient pigments.[15] By using eggs from living birds as controls to optimize their analytical techniques, along with a large variety of fossil eggs from both the Peabody Museum of Natural History at Yale University and the American Museum of Natural History, they were able to show that dinosaur eggs were indeed colored. Of the 14 fossil dinosaur eggs they examined, most contained the same pigments as living birds. Given the various ages of the dinosaur eggs, the researchers were able to estimate that egg pigmentation first evolved in nonavian dinosaurs, perhaps about 150 million years ago. Colored eggshells appear to be a characteristic of organisms that have open (vs. buried) nests. Pigmentation, functioning as camouflage, protects against predation while biomineralized shells protect against dehydration and provide the structural support required for brooding.

Dinosaur Vision

If dinosaurs had brightly colored feathers and colored eggs, it is reasonable to assume that they could see colors. But is there any physical evidence of color vision in dinosaurs? Birds are known to have excellent color vision—better than ours in fact.

The shortest wavelengths of light that humans can see, the violets, are about 400 nanometers in length, while birds can see the colors of much shorter ultraviolet wavelengths. Is there any way of knowing what colors the dinosaur eye could see? Recent research by Gengo Tanaka and his colleagues has shown that at least one ancient theropod appeared to have had vision similar to that of modern birds.[16] The dinosaur was a 120-million-year-old member of a group of ancient bird-like theropods that had both teeth and wings with clawed "fingers." When a small piece of the eye of this fossil was removed and examined with a scanning electron microscope, not only were cone cells visible, they were identified as containing permineralized oil droplets. (I must admit that when I first read Tanaka's work, I read "oil droplets," not "permineralized . . . oil droplets," and visions of an array of preserved cone-specific ancient pigments danced in my head. Unfortunately, minerals had replaced the pigments.)

Although color vision is provided by the cone cells of the retina, these structures differ dramatically between mammals and birds. Each bird cone cell has a large droplet of oil placed so as to separate the light-responsive proteins at the cell's apex from the neuronal synapse at the base of the cell (human cones have nothing resembling an oil droplet). The oil droplets, which in living organisms are rich in carotene-related pigments, absorb different wavelengths of light. There are five different types of cone cells, each containing an oil droplet with a different carotene-like pigment, and each transmitting its own unique and narrow spectrum of light. As a result, the different types of cones have differently colored oil droplets.

By measuring nearly a thousand oil droplets in cone cells from the eyes of extant sparrows, chickens, and quail, Tanaka determined that the diameter of the oil droplets correlates with the type of pigment in the cone cells. In the quail eye, the smallest droplets transmit ultraviolet light; droplets intermediate in size transmit blue, yellow, and green light; and the largest droplets transmit red light. Tanaka also found that diameters of oil droplets in the fossil dinosaur eye varied over tenfold. Does this indicate that dinosaurs could see an even wider array of colors than most modern birds?

When interpreting a finding like this, the first thing that comes to a paleobiologist's mind is the effect of death, burial, preservation, and compression on the size of the fossilized oil droplets. The impossible task of trying to replicate the processes of fossilization exactly as they occurred in nature has been the bane of many a researcher; what "control" experiments does one use when studying the formation of deep time fossils? Some scientists have gone to great lengths in their attempt to reproduce degradation over time. For instance, Mary Schweitzer's laboratory at North Carolina State University buried bird feathers in sand for 17 years, then dug them up for analysis.[17] This is not, however, an experiment that an assistant professor working towards tenure can undertake. Given the uncertainties, Tanaka and his colleagues were laudably conservative in their interpretation and concluded only that the dinosaur could detect a wide range of colors, including perhaps, the colors of its mate.

As the science of ancient biomolecules advances and documents additional dinosaur color patterns, the construction of three-dimensional models of dinosaurs will incorporate multicolored feathers. Both paleobiologists and moviegoers will have access to more accurate (and exciting) representations of what dinosaurs were actually like. While many of these future studies will rely on pigments, there is more than one way to skin the proverbial cat. Making a unique observation, Roy

Wogelius and his colleagues noticed that the distribution of copper in the fossil bird *Confuciusornis sanctus*, collected in China, coincided with those portions of the fossil composed of feathers.[18] Can copper be used as a proxy to identify fossil feathers? It's an interesting question, and one that leads us to an examination of a relatively unappreciated group of ancient biomolecules: biometals.

6

ANCIENT BIOMETALS

We don't usually think of metals as biomolecules. We commonly understand biomolecules to be large structures with many interconnected atoms of carbon, hydrogen, oxygen, and various other types of elements. A single molecule of hemoglobin, for instance, has 8,456 atoms, not counting its four atoms of iron. But copper sulfide (CuS), with just two atoms, is also a molecule. Individual atoms of most metals are rarely found free—that is, they are usually bound to something else. And when the something else is part of a living organism, the metal becomes a biometal.

The percentage of Earth's metals bound up in biomolecules, in living organisms, is vanishingly small. But without these biometals, life would not exist; biometals are critical components of respiratory pigments like hemoglobin (with its iron), the blue hemocyanins found in many arthropods and mollusks (with its copper), and photosynthetic molecules such as chlorophyll (with its magnesium). There are also thousands of proteins that contain biometals. For example, tyrosinase, the enzyme that mediates the first reaction in both sclerotization and the synthesis of melanin, contains two atoms of copper that are essential for its activity. Such proteins are called metalloproteins and their study is called metallomics.

For ancient biometals to play a useful role in the study of ancient life, the metal must be found in situ. In other words, it must be restricted to one specific part or structure within the fossil—the same structure in which it was found when the organism was alive. If iron had been found in the head, wings, and legs of the blood-engorged mosquito at the same levels as that found in the distended abdomen, it would have been impossible to conclude that remnants of blood had been preserved. Mere detection of diffuse unlocalized metal in a rock is of no value (unless you're the Rio Tinto Mining Company). Thankfully, however, the fossil need not contain very much of it. Modern analytical techniques can detect exceedingly small amounts of biometals, often as little as a few parts per million. And such technology is readily available at the Smithsonian.

Metal Mandibles

The National Museum of Natural History, with nearly 600 employees and as many volunteers, houses a wide array of scientific disciplines under a single roof. On any given day, there are numerous departmental seminars and journal clubs, discussing matters from anthropology to zoology, available to everyone in the building. Since one never knows where the next good idea will come from, I consider it good practice to take in seminars on topics seemingly peripheral to my own research interests, as do many scientists at the museum.

As a result of this habit, I found myself attending a seminar about trap-jaw ants given by Fredrick Larabee, at the time a graduate student at the University of Illinois, Urbana-Champaign. These predators hold their large jaws (mandibles) wide open while foraging, and when they sense contact through their jaw's sensory hairs, they snap them shut with incredible speed, as fast as 210 feet per second. (The Dracula ant *Mystrium camillae* holds the speed record at 295 feet/second.)

During his seminar, Larabee mentioned as an aside that some ants have mandibles that are hardened by the incorporation of zinc. As soon as he said this, the proverbial light came on: perhaps zinc, a metal, would have a better chance of being preserved within a deep time fossil than a more labile organic biomolecule. Could we find zinc within the jaw of an ancient insect?

A review of the literature revealed that scientists had known about the metal-based hardening of insect mandibles for decades. And not just mandibles; claws and the tips of ovipositors (egg-laying structures that, in large wasps, have evolved into stingers) are also strengthened by the incorporation of zinc. A search of the museum's database revealed a number of fossil insects with beautifully preserved mandibles. I selected a fossil rove beetle (family Staphylinidae) and made a trip to the Entomology Department to borrow some pinned specimens of extant rove beetles to serve as controls. As I had done in the identification of iron in the abdomen of the fossil blood-engorged mosquito, I contacted Tim Rose in the museum's Mineral Sciences Department. I hoped that Tim's technical expertise in the identification and quantitation of individual elements would allow us to determine whether or not zinc had survived, in situ, for 46 million years. Using a nondestructive X-ray beam that, even at relatively high energies, penetrates to a depth of only several microns, Tim analyzed the extant specimens, as well as their fossil ancestor.

As anticipated, zinc was found in the mandibles of several of the extant specimens. The terminal tip of the mandible, the bicuspid molar area (the sharp "teeth" at its base), and the cutting edge of the mandible were the main sites of zinc incorporation. While Tim performed his analysis, I could not help thinking

FIGURE 6.1. The jaws of a 46-million-year-old fossil beetle (left) still contain the metal zinc (right) that was used to harden the edges of its mandibles.

about Jaws, the Bond villain from *Moonraker* and *The Spy Who Loved Me*, whose teeth were capped with steel. When the fossil beetle was analyzed, we were excited to find a zinc distribution similar to what we found in the extant beetle. The metal was present along the sharp inner edge of the mandible, as well as in the teeth at the base of the jaw.[1] The zinc hardening agent preserved in a 46-million-year-old insect mandible was a new and valuable example of an ancient biomolecule—in this case a biometal—with a known function.

Scientific discoveries nearly always lead to more questions. What holds the zinc in place at the tips and edges of the mandibles? Is there a mandible-specific zinc-binding protein? Melanin is known to bind a number of various metals; could the pigment perform yet another function and provide a scaffolding for jaw-hardening metals? Or might chitin, which is known to bind to a variety of metals, be responsible for the preservation of the fossil mandible's zinc? Unfortunately, almost nothing is known about localization of zinc in the mandibles of living insects, let alone 46-million-year-old fossils. It would take a great deal of additional work to answer the question, and the mandibles of the fossil beetle would have to be removed, ground up, and extracted—a practice that almost no scientist would endorse. (During a seminar at the Calvert Marine Museum, I asked for a show of hands: Who supported destructive sampling that would allow us to conduct immediate analysis? And who thought we should wait until new, nondestructive techniques were developed? Not a single hand was raised in support of an attempt at immediate analysis.)

Beyond shedding light on the past, work on the ancient mandibular zinc may have practical applications in today's world. Over the last billion years or so, nature has produced large numbers of very unusual proteins that, even with modern science's sophisticated knowledge of protein chemistry, we would never think of designing ourselves. Several laboratories, such as those of Markus J. Buehler at the Massachusetts Institute of Technology, are presently attempting to use and even improve on nature's blueprints, with the aim of producing artificial biopolymers for commercial purposes. Buehler's lab has, for instance, investigated a zinc-binding protein in the jaws of *Nereis*, a marine worm. They were able to create a synthetic version of the protein and show that aggregates of the protein became more compact in the presence of zinc.[2] Our understanding of the many ways that metals interact with proteins may have profound medical applications.

Buehler's lab has also investigated one of the most interesting natural fibers on earth, spider silk. In some ways stronger than steel, spider silk has long attracted attention. Nearly 20 years ago, a Canadian biotechnology company called Nexia Biotechnologies received a huge amount of press coverage when they announced that they were able to produce recombinant spider silk in the milk of two transgenic goats named Sugar and Spice. The company eventually went bankrupt: they could make the silk protein, but they couldn't spin it into a commercial product because their scientists were unable to figure out how to duplicate the silk assembly plant found in a spider's behind. Over the past two decades, however, progress has been made on this front. Silk can now be synthesized in both bacteria and plants. In fact, a company in Japan called Spiber has developed a commercial process by which massive amounts of recombinant silk fibrils can be manufactured.[3] Companies have also become more realistic about the application of transgenic silk. The silk-based bulletproof vests and bridges supported by silk cables pictured in 1960s comic books have given way to silk casings for plant-based sausages.

Metallic and Nonmetallic Markers

The localization of atoms of metals and other elements in deep time fossils can provide many different kinds of important information about ancient organisms: phylogenetic, behavioral, physiological, evolutionary, and anatomical. Visualization of phosphorus in vertebrate fossils, for example, can provide a highly detailed map of their bony skeletons. One study by Roy Wogelius's team, at the Interdisciplinary Centre for Ancient Life at the University of Manchester in the United Kingdom, involved a unique 50-million-year-old reptile collected from the Green River Formation in Colorado.[4] The specimen consisted of nothing except the skin; no readily visible bones had been preserved—the exact opposite of what normally happens during most cases of fossilization. Although the entire organism's skin was intact, and distinct scale-like patterns could be seen on its surface, there was little morphological information on which a taxonomic identification could be made. That is, until the fossil was analyzed for the element phosphorous, an easily detected marker for bone, which is made largely of calcium phosphate. Analysis confirmed that the bones of the reptile's skeleton had been dissolved prior to their fossilization—no traces of any of the lizard's bones were left. What was visible, however, were two sets of 14 to 16 teeth, preserved because tooth enamel is more resistant to dissolution than bone. The dentition pattern indicated a relatively long and narrow jaw. Based on this observation, Wogelius and his colleagues were able to assign the lizard to the reptilian family Shinisauridae, a family that contains only a single living species, the Chinese crocodile lizard that lives today in China and Vietnam.

The use of the element phosphorous to find teeth and bones has become a standard application of ancient biomolecule research. However, it represents only a small fraction of the potential of both metallic and nonmetallic elements to elucidate the anatomy of fossil organisms. In a recent study by Maria McNamara and her colleagues, the distribution of the elements zinc, copper, iron, sulfur, manganese, and titanium were determined in both extant animals and fossils.[5] They demonstrated, for example, that in extant birds and mammals, the spleen was enriched in iron relative to the liver. And when the tadpole *Pelophylax pueyoi* from the Miocene Libros fossil site in Spain was examined, some but not all of the elements showed a degree of tissue-specific localization. For example, zinc was concentrated in the eye, notochord, and tail muscles.

Anatomy is, moreover, only one of the many aspects of life that ancient biometals and related elements are helping us understand. In a recent study by David Terril and his colleagues at the University of Calgary, localization of a different element, sulfur, has been used to shed light on the evolution of some of the oldest known vertebrate fossils.[6] Conodonts are infamous among paleobiologists because, although first discovered in

1856, we still have a relatively poor understanding of their phylogenetic relationships to other organisms. They look like tiny rows of teeth; most are less than a few millimeters long. They are quite common as fossils go—almost 200 different genera have been named, and they lived over a span of 300 million years (roughly 525 to 210 million years ago). They finally disappeared during the huge extinction event that defined the boundary between the Triassic and Jurassic periods. Paleobiologists value them primarily because different species of conodonts serve as markers for rocks of different geological ages. Yet we still don't know exactly what they are.

They are vertebrates, everyone agrees on that, especially since Derek Briggs and his colleagues discovered a fossil near Edinburgh, Scotland, about 40 years ago that consisted of conodont-like teeth and the tiny eel-like organism to which they belonged.[7] But even today, there is disagreement about the placement of the conodonts within the evolutionary tree. Lampreys and hagfish lack jaws, and instead of teeth, they have knob-like structures made of the protein keratin, the same type of protein responsible for horns, hooves, fingernails, hair, and feathers. Conodonts also lacked a jaw, but their teeth have been shown to consist, in large part, of a calcium phosphate mineral similar to that in our teeth and all other jawed vertebrates. There is no consensus regarding the evolutionary relationships of these several different groups; one can find both DNA sequence data (from living species) and morphological data that support competing theories.

Terril examined conodonts for the presence of sulfur. His reasoning was straightforward: the protein keratin contains lots of sulfur, about 4 percent of its weight. (This is why, for instance, the burning of hair produces such a nasty odor: vertebrate hair is made of keratin, and when it burns, we smell the oxidation of sulfur.) Terril was looking for keratin, but he knew that, at 260 and 465 million years of age, the conodont teeth that he was studying had no preserved protein. He theorized that while the protein may not have been preserved, perhaps the sulfur had stuck around, just as the zinc had been preserved in the fossil beetle jaw. He decided to use sulfur as a proxy for the presence of ancient keratin. And he found it; about 0.2 percent of the conodont fossils he examined consisted of sulfur. Terril made two additional observations that were significant to the discussion of conodont phylogeny. He first determined that the sulfur was not present simply as a mineral such as pyrite (iron sulfide); much of it appeared to be bound to organic material. He was then able to localize the sulfur within the conodont tooth. Microscopy demonstrated the presence of numerous ring-like layers within the teeth, an indication of the growth of the calcium phosphate-containing portions of the teeth. In the middle of each tooth, he found the sulfur concentrated between the youngest layers. Terril theorized that the sulfur was a remnant of a keratin framework at the center of each young conodont tooth. The possible presence of the keratin, as suggested via the presence of sulfur, led Terril to conclude that, despite the presence of mineralized teeth, conodonts were more closely related to lampreys and hagfish than jawed vertebrates.

Biometals are also helping us understand the color patterns of ancient life. For example, studies from Roy Wogelius's laboratory have allowed us to reconstruct the color patterns of extinct early birds and feathered dinosaurs. Wogelius and his colleagues noticed that the distribution of copper in the fossil bird *Confuciusornis sanctus*, collected in China, coincided with those portions of the fossil composed of feathers.[8] When other fossil bird feathers were examined, similar results were obtained. As it turns out, the polymeric pigment melanin can bind numerous different

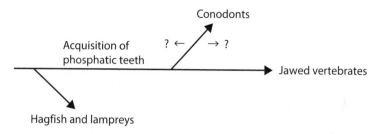

FIGURE 6.2. The presence of teeth containing both phosphorous and sulfur suggested a new interpretation of the relationships between hagfish, lampreys, and conodonts.

metals, including zinc, calcium, and copper. In fact, the percentage of copper in the cross-linked melanin polymer can be as much as 6 percent of its total weight, a huge amount—for comparison, iron makes up only 0.33 percent of the weight of hemoglobin. Might metals act as a proxy for the presence of melanin? It appears to depend on the context. For example, although plants do not make melanin, Wogelius's group demonstrated that the distribution of copper in a fossil leaf provides an image that is a near perfect replica of the leaf. Within the context of a feathered dinosaur, however, ancient copper appears to accurately reveal areas, but not necessarily the colors, of melanin-based pigmentation.

Wogelius and his colleagues followed up their initial observations about copper in fossil leaves by initiating more detailed studies of the metallomes of plants.[9] They first analyzed the distribution of numerous metals, including copper, nickel, and zinc, in the leaves of an extant American sweetgum tree. Previous studies by other laboratories had shown that several species of willow trees accumulate metals at the very tips of the serrated edges of the tree's leaves, and Wogelius's study found a similar situation in sweetgum trees: they contained very high concentrations of zinc, copper, manganese, and nickel at the tips of their leaves. Were the plants trying to reinforce and strengthen the fragile and exposed tips of their leaves' margins with metals? In this case, no. Plants have evolved mechanisms to rid themselves of excess toxic metals by isolating and sequestering them in the serrated tips of the leaves—in some cases, they actually excrete them. Wogelius went on to show that this same phenomenon existed 50 million years ago. He and his colleagues localized high concentrations of copper at the tips of the serrated margin of leaves of the fossil *Platanus wyomingensis* (an extinct species of sycamore tree), collected from the Green River Formation in Wyoming.

In a different fossil leaf, which had been the victim of extensive herbivory by the larva of a moth, the presence of ancient biometals was able to document its demise. The insect had eaten the soft tissue that filled the spaces between the many veins of the leaf in a process that entomologists call skeletonization. As for all organisms, the more one eats the more one defecates, and this particular larva had a healthy appetite. The larva left behind three curlicue-shaped trails of frass, the polite scientific term for larval insect excrement. As all gardeners know, caterpillars have ravenous appetites, and it doesn't take them long to consume an entire leaf. Wogelius and his team found that the leaf's metals had been concentrated in the frass. The trails of the excrement easily stood out against the skeletonized portions of the leaf when analyzed for copper and other metals.

Frass, having been processed by the intestine of the caterpillar, undoubtedly contains remnants of DNA and proteins from the insect's intestinal cells and secretions. Although the cloning of an extinct butterfly from fossilized caterpillar poop may not be as sexy as the cloning of dinosaurs from the fossil of a blood-engorged mosquito, it is, in theory, possible, even if it is unlikely that the DNA would survive the long journey to the present intact. But what if we lowered our sights and examined fossil frass that is relatively young, and if we made protein rather than the more labile DNA our target: would it be possible to identify the type of insect that produced the frass? Still a tall task, but proteins are a major focus of ancient biomolecule research, and once we consider what researchers in the field of ancient proteins have already accomplished, our proposed experiment will perhaps not look so unreasonable.

7

PROTEINS AND PROTEOMES

In 1935, the German paleontologist Gustav von Koenigswald walked from drugstore to drugstore in Hong Kong, so that he could examine the bowls of "dragon teeth" that the stores sold for use in folk medicine. It wasn't exactly fieldwork, but it turned out to be very productive. At one store, von Koenigswald purchased a tooth unlike any he had ever seen. It was from a primate, an herbivore, but it was unique both for its size—it was huge—and its morphology. He subsequently used the tooth as the basis for his description of a new hominid, *Gigantopithecus blacki*.[1] Since that time, numerous other teeth and a few partial jawbones have been found, enough to calculate the size of the ape: around 10 feet tall and 600 pounds.

The great apes today consist of four genera: humans, chimps, gorillas, and orangutans. Both morphological and genetic studies provide us with a phylogenetic tree that firmly details the evolutionary relationships (or phylogenies) of these four groups. But where does *Gigantopithecus* fit in? The scant morphological data offer no clues, and there is no ancient DNA available for molecular phylogenetic analyses. To solve this mystery, scientists would have to rely upon a type of molecular analysis that could be conducted without DNA. A little history will explain how this might be done.

In the late 1950s and early 1960s, scientists developed technology to sequence proteins; that is, to determine which of the 20 different protein subunits, or amino acids, are in any given position in the beads-on-a-string structure of proteins. As increasing numbers of proteins were sequenced, scientists began to notice slight differences in the sequences of a given protein from different animals. In 1961, Emile Zuckerland, a postdoctoral fellow in Linus Pauling's lab at the California Institute of Technology, realized that these differences could enable scientists to create a "molecular clock."[2] In essence, the molecular clock is a measure of the "evolution" of a protein—the sequence of changes that occur over time due to mutations that alter the DNA sequence of the gene that encodes the protein. Zuckerland proposed that the same data, from one protein, could be used to document the evolution of the organism from which the protein sequence came: the rate at which mutations in proteins occur was thought to be constant, so the more sequence changes its proteins had relative to those in more recently evolved species, the longer the organism must have been around. For example, while sequences of hemoglobin from a cutthroat trout and a rainbow trout may be very similar, if not identical, both differ dramatically from sequences of the living but very primitive and evolutionarily ancient coelacanth from the Indian Ocean. Ideally, of course, it is best to compare sequences of the genetic material itself, DNA, as sometimes a mutation in a gene will not lead to a change in the amino acid sequence of the protein it encodes. But when DNA is not available and protein is, the latter will do, as its sequence is a direct reflection of the encoding DNA sequence. Zuckerland and Pauling went on to demonstrate, not without criticism, that molecular clocks provided data that agreed with conventional morphology-based phylogenetic data (i.e., evolutionary trees).

An examination of Pauling's personal notes provides a fascinating look at the controversy that accompanied the development of the molecular clock.[3] The famous evolutionary biologist C. H. Waddington had criticized the idea on the grounds that "Evolution of the lion does not involve what sort of hemoglobin the lion had in its blood, but rather, does the lion catch the antelope or not? This depends on thousands of genes, not a single gene." Responding to Waddington and other leading scientists of the time, including the intellectual giants J.B.S. Haldane and Sewall Wright, Pauling noted that they did not appreciate the concept that changes in a single gene, and the protein that it encoded, could have a huge effect on an organism. As he recorded in his notes, "the lion will not catch the antelope if it has a poor hemoglobin gene."

Today, the idea of the molecular clock—that the number of substitutions in the gene that encodes a protein's amino acid sequence can be used as a proxy for the age of the evolutionary origin of an organism—is, with caveats, readily accepted. Variations in the gene's sequence can provide information about how closely related one species is to another, as well as about changes in protein functions and protein-dependent physiology over deep time. For example, a recent application of the molecular clock examined the evolution of myosin, a major protein of muscle. Myosin gene sequences of seven nonhuman primates and *Homo sapiens* revealed a single mutation that was unique to modern humans.[4] Calculated to have occurred approximately 2.4 million years ago, the mutation inactivated one of several myosin-encoding genes. Loss of this unique form of myosin correlated with loss of the powerful masticatory muscles that characterized the primate ancestors of the genus *Homo*, as well as *Homo erectus*. Scientists speculate that the resulting "gracilization" of the human skull led directly to an increase in the brain capacity of *Homo sapiens*.

Molecular divergence determinations are not, however, as simple as initially proposed. Different genes within the same organism mutate at different rates, and amino acids at different positions within the same protein undergo substitutions at different rates. Moreover, mutation rates may have changed over geological time. Despite these complications, numerous improvements to the molecular clock have been made over the years, including the development of new statistical algorithms, and the field of molecular phylogenetics is now well established. In fact, at one point, over-enthusiastic advocates of the molecular approach concluded that fossils were no longer needed. Molecular techniques, by themselves, were thought to be perfectly capable of constructing phylogenies and divergence times for all living species. However, such bold claims were abandoned when DNA sequence data predicted that a certain group of insects first evolved 50 million years ago, yet a fossil from that group was found in 85-million-year-old shale. Today, it is recognized that both types of data are essential to our ability to construct accurate phylogenies, and a Fossil Calibration Database that consists of well-vetted fossils is used to establish an absolute minimum age for the divergence of specific taxa from their sister lineages.

To apply the molecular clock to the evolution of the *Gigantopithecus blacki*, scientists would first need to locate ancient proteins from the mysterious creature. In 2006, Wei Wang, an anthropologist at the Institute of Cultural Heritage at Shandong University, would make this critical discovery in the valley of the Bubing Basin, just north of the Vietnamese border. The valley is bordered by towering limestone rock formations, and near the top of one of the formations, 250 feet above the valley floor, a small entrance leads to a 30-foot-long cave. The soil on the cave's floor, which was orange and four feet deep, had been

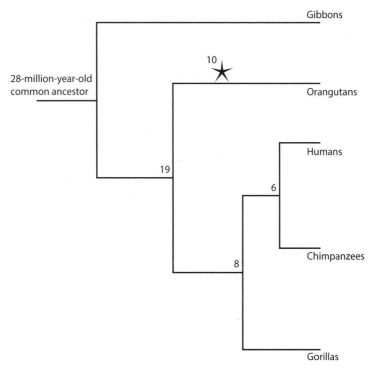

FIGURE 7.1. A phylogenetic tree of the great apes. The asterisk indicates the divergence of *Gigantopithecus blacki* from the ancestor of the Orangutans. The numbers indicate the divergence dates, in millions of years, of the various lineages.

found to contain the ancient bones of numerous animals that dated to about 2 million years ago. Wang slowly and meticulously removed half-inch-thick layers of the soil; when he reached a depth of about four inches, he uncovered two beautiful molars. White and shiny, they looked like they had only recently been separated from an animal's jaw. In fact, they were 1.9-million-year-old teeth of *Gigantopithecus blacki*.[5]

Extracting the necessary biological material from these fossils would take nearly another decade, but working with Frido Welker and Enrico Cappellini of the University of Copenhagen,

Wang was eventually able to extract partial sequences of six different ancient proteins from the tooth's enamel.[6] When the ancient protein sequences were compared to those of other hominid species, molecular clock-based phylogenetic analysis indicated that *Gigantopithecus blacki* was indeed a hominid. It had evolved from an ancestor that also gave rise to the orangutan about 10 million years ago. In the parlance of phylogenetics, the genus *Gigantopithecus* is sister to the orangutans (*Pongo*), just like *Homo* is sister to *Pan*, the chimpanzees.

The Power of Ancient Proteins

As the case of *Gigantopithecus blacki* indicates, ancient proteins have become a game-changing tool for anthropologists and paleontologists. Ancient proteins can tell us when an organism first appeared on Earth and to what other organisms it is most closely related. Establishing the evolutionary relationships of otherwise poorly documented long-extinct organisms can shed significant light on the broader development of life on Earth, so it is unsurprising that much of the ancient protein research to date has concentrated on phylogenetics. One study, in particular, exemplifies the degree to which ancient protein sequence-based phylogenetics can force us to revise our beliefs about the past.

The extinct genus *Plesiorycteropus* was first collected in western Madagascar in the late 1800s. For the next 200 years, it was considered to be closely related to aardvarks, based on the animal's unique digging-type feet. Then, about a quarter of a century ago, it was suggested that the clawed appendages may have been acquired as the result of convergent evolution. Because the specimens exhibited no other unique morphologies, they did not fit within any other existing taxa, so an entire new order,

Bibymalagasia, was created to accommodate the organism. But when a specimen was taken off a museum's shelf and analyzed for ancient proteins, it was found to have preserved collagens. Sequence data from the collagen proteins readily identified the *Plesiorycteropus* as belonging to the well-established order Tenrecoidea (golden moles and tenrecs, the latter a highly diverse group of small insectivorous mammals).[7] The order Bibymalagasia was neither valid nor needed.

There are tens of thousands of bone specimens in museums throughout the world that are either unidentified or misidentified, many of which contain an invaluable reservoir of ancient proteins that will aid in their identification. However, phylogenetics is only one area in which ancient proteins promise to advance scientific understanding. Compared to pigments, for example, protein sequences have a far greater potential to reveal the behaviors and physiologies of long-extinct organisms. In some ways, ancient proteins are even superior to ancient DNA: proteins appear to be more stable, and they are preserved far longer into deep time. One potentially negative aspect of DNA is that many genes are expressed (translated into proteins) only at certain times or under certain conditions. A recovered ancient DNA sequence would indicate only that the extinct organism had the ability to make a certain protein; it doesn't mean that the protein had been synthesized or was functioning at the time of the animal's death. On the other hand, if an ancient protein sequence is recovered from a fossil, we can be sure that it was a functional component of the organism and contributed to the animal's activity, health, and appearance.

Consider, for example, what more we can learn about the giant hominid *Gigantopithecus blacki* from the recovery of its fossil tooth. There are 2,479 different types of proteins in human bone tissue, and although most are present in vanishingly small amounts, together they make up about 30 percent of the weight of bone tissue.[8] Scientists were surprised, however, when they isolated a protein called fetuin-A from the enamel of the hominid's tooth. Fetuin-A is made in numerous tissues in the mammalian body, but it had never been found in tooth enamel. In other tissues, fetuin-A participates in the formation of calcium-protein complexes and operates as a kind of mineral chaperone. Given the exceedingly thick enamel crowns of the molar of *Gigantopithecus blacki,* Welker and Cappellini were willing to propose a new function for the protein: mediating the thickening of the tooth's enamel.

Perhaps even more striking, proteins in the enamel of the tooth revealed the sex of *Gigantopithecus.* The protein amelogenin comes in two varieties, one with a male-specific sequence and one with a female-specific sequence. Scientists were able to recover amelogenin from the enamel of *Gigantopithecus blacki* and determine that she was, well, female. In another study, scientists recovered ancient enamel proteins from a 1.77-million-year-old extinct relative of the rhinoceroses and determined that the specimen was a male.[9]

Ancient Collagens

The scientists who sequenced ancient proteins from the teeth of *Gigantopithecus* referred to the recovered proteins collectively as an "enamel proteome." A proteome is to proteins what a genome is to genes: a compilation of proteins (or genes) of an organism. One major difference between the two, however, is that while the entire compilation of genes is conveniently concentrated in any one of the trillions of tiny cells in our body, the 20,000+ different proteins (of humans) are scattered throughout the body in a variable manner; there are brain-specific proteins, liver-specific proteins, eye-specific proteins, etc. While it may be possible to grind up a mosquito and extract all of its proteins (a mosquito proteome), this would be impossible to do with, say, an elephant—it would be much easier to produce an elephant genome and predict what proteins should exist based on gene sequences. Proteomes are therefore usually subsets of an organism's proteins: liveromes, brainomes, canceromes, etc.

The largest nonhuman ancient proteome, isolated from bone tissue of a woolly mammoth (*Mammuthus primigenius*) found frozen in 43,000-year-old Siberian permafrost, consists of 126 proteins. This might, at first, sound like a bonanza. Recall, however, that this represents only about 5 percent of the total number of different proteins that are found in bone tissue.[10] To add to the bad news, recovery of complete sequences is exceedingly rare. On average, for the 126 proteins in the *Mammuthus primigenius* proteome, only 14 percent of each sequence was determined; the most complete sequence, a type of collagen, was about 72 percent complete. Deep time extracts a toll. Fortunately, for phylogenetic studies, complete sequences are not required. In most cases, enough mutations are spread throughout the protein sequence to ensure that even a fragment will contain mutations and allow for comparisons with related extinct and extant organisms.

Until recently, the oldest proteome, with 73 proteins, was from an approximately 670,000-year-old horse found at the permafrost Thistle Creek site in the Yukon of Alaska.[11] That record has now been surpassed by the enamel proteomes of the 1.77-million-year-old rhinoceros and the 1.9-million-year-old *Gigantopithecus blacki*. Again, however, there were severe costs associated with going back further in time. In both cases, bone samples were analyzed and found to be devoid of ancient protein. And although the ancient enamel revealed important information about the creatures, "Enamelomes" from even extant teeth contain far fewer proteins than bone tissue. Most ancient proteomes have been obtained from bones of relatively younger megafauna. They include a 120,000-year-old giant-horned bison (*Bison latifrons*) from the Snowmass site in Colorado, a 40,000-year-old bovine dredged up from the North Sea, and an 11,000-year-old specimen of *Mammuthus columbi*, also from Colorado.[12]

In the bison proteome, more than a third of the proteins were collagens; in the horse from Alaska, 10 different types of collagens were obtained. Similarly, most of the sequences obtained from the 11,000-year-old mammoth were from collagens. Obviously, there is more to an organism than collagen, and the recovery of the sequences of noncollagenous proteins will be essential if ancient proteomes are to become a more valuable tool in our quest to understand ancient life. Nevertheless, ancient collagens currently are, and will continue to be, the focus of much of ancient protein research.

Collagen is, after all, one of the most abundant proteins on the planet. It makes up about 28 percent of bone, and it is the major protein in skin, blood vessels, tendons, and ligaments. In the human body, collagens account for over 25 percent of the cumulative mass of proteins, and there are 28 different types of collagens found in vertebrates alone. Thought to have evolved more than 600 million years ago, collagen is present in all multicellular organisms. The basket-shaped fibrous skeleton of *Vauxia gracilenta*, the chitinous sponge from the Burgess Shale, is assumed to have been, in part, composed of collagen-like proteins called spongins.[13]

Given that paleobiologists encounter collagen far more frequently than any other ancient protein, they have learned to take advantage of its seemingly ubiquitous preservation. In fact, newly developed methodologies have made the role of ancient collagen even more prominent. A new technique, developed by Michael Buckley, currently head of the Ancient Biomolecules Laboratory at the University of Manchester, relies on the ability of enzymes to cleave collagen at very specific places.[14] For example, the enzyme trypsin, which functions in our small intestine to digest proteins, will cut a protein only where there is an amino acid that contains a nitrogen in a form similar to ammonia. The numerous fragments that result from the digestion can be separated, based on their mass, and portrayed as a spectrum or series of peaks (fragments) on a graph. Different kinds of collagen will have sequences with different numbers and placements of the amino acids recognized by the enzyme and will therefore provide different spectra. And because the sequences of collagens from different animals differ, they too will produce spectra unique to their species. Buckley and his colleagues analyzed the collagen proteins from a large number of different species of living animals and created a reference library of validated species-specific collagen protein fragmentation patterns, or fingerprints, that everyone could use. With this new technique—called collagen fingerprinting—scientists are able to identify the source of an ancient collagen without needing to sequence it.

This methodology represents a major advance, as collagen fingerprinting is easier, faster, and less expensive than sequencing, and can be applied to collagens in everything from bones to eggshells to mummies. Using automated search algorithms that compare the collagen fingerprints of fossils to those in his reference libraries, Buckley can quickly and easily find matches and identify unknown fossil bones to genus and even species. The technique has been adopted by archaeologists, anthropologists, and paleobiologists around the world, including those who do their fieldwork in the northernmost reaches of Canada.

Ellesmere Island, the third largest of the thousands of islands that constitute the Canadian Arctic Archipelago, lies just off the northwest tip of Greenland at a latitude of 78 degrees north (for context, Point Barrow, the northern-most point in Alaska is about 470 miles closer to the equator). The island goes without sun for six months of the year and is perhaps the last place on Earth one would expect to find a camel. The genus *Camelus* consists of three living species: the Bactrian and the distinct wild Bactrian camels of Asia, both of which have two humps, and the Arabian (dromedary) camel with its single hump. This genus belongs to the family Camelidae, which also includes the alpacas, guanacos, and llamas of South America. Despite its current distribution, camelids originated in North America approximately 45 million years ago and diversified into more than 53 different species. Like the horse, genus *Equus*, ancestors of *Camelus* dispersed to Asia via migration across a Beringia land bridge and eventually disappeared from the New World. But not so long ago; the leg bone of a specimen of the extinct genus *Camelops*, discovered in a gravel pit in Alberta, Canada, was recently determined to be just 11,280 years old.[15]

Much of the Arctic, in both Asia and North America, is thought to have been covered by coniferous forests during the late Miocene, 11.6 to 5.3 million years ago. Species of the extinct genus *Paracamelus,* thought to be direct ancestors of modern camels, lived in these forests as long as 7 million years ago. Scientists have even proposed that the iconic hump of modern camels evolved in *Paracamelus* not, as I remember being taught when I was a child, as a reservoir of water to aid in desert travel, but rather as a reservoir of energy-packed fat for use during the very long and cold winters. A distinct population of *Paracamelus* moved even farther north to a large land mass that only later, with rising sea levels, became the Canadian Arctic Archipelago.

About 3.5 million years ago, one of these animals died, and its skeleton was widely dispersed, very widely dispersed: its bones were eventually collected over a period of five years on Ellesmere Island by Natalia Rybczynski of the Canadian Museum of Nature. Rybczynski and her colleagues collected about 30 small fragments, some of which were only a half inch in length. Given their small size, standard morphological analyses could not be applied—it was impossible to identify the bones as those of a camelid. In fact, there was no certainty that the bones even belonged to a single specimen. Fortunately, the bones contained ancient collagen.[16]

The 30 bone fragments of the Ellesmere Island specimen were analyzed by Buckley with the collagen fingerprinting technique. Against all odds, each was determined to be from a *Paracamelus*-related organism with striking sequence resemblance to modern Arabian camels. The fact that all 30 fragments, accumulated during three different collecting trips, were all from the same species, and probably from the same animal, was simply incredible. The results of the fingerprinting analyses allowed the bone fragments themselves to be fitted together like a jigsaw puzzle, forming the lateral surface of the animal's right tibia. Based on that reconstruction, the Ellesmere Island camel was estimated to be much larger than modern camels, similar in size to giant extinct camels such as *Paracamelus gigas,* which weighed as much as 3,000 pounds.[17]

As it turns out, producing collagen fingerprints for 30 samples is a trivial task compared to some of the work Buckley and his colleagues have undertaken. In another study, they examined bone fragments from a cave in the East Midlands of England. Caves were, of course, convenient shelters for humans, as well as for large numbers of other mammals, and the 100-foot-long Pin Hole cave was no exception. Narrow, as little as three feet

wide in places, it expanded internally to a comfy chamber with
evidence of inhabitation by both Neandertals and, much later,
modern humans. But what else? As is often the case, the pieces
of bone collected from Pin Hole were fragmented to such an
extent that they could not provide meaningful taxonomic information; the researchers turned to collagen fingerprinting.
Fingerprints for 12,317 bone fragments were produced, and
bone collagens from nearly 20 different mammalian genera
were identified, with reindeer by far the most common. Also
present were hyenas, lions, woolly rhinoceroses, bears, and arctic foxes—some perhaps the prey of the hominins.[18] Caves that
served as homes for humans, other primates, and an array of
other organisms provide anthropologists with a windfall of invaluable information. Their potential to widen our window to
the lives of these ancient communities has increased exponentially with the advent of ancient protein technologies.

Collagen fingerprinting has been widely adopted by paleobiologists and anthropologists. But what about ancient noncollagenous proteins? Do collagens dominate simply because they
are more common, or is it because they are more resistant to
degradation? What does it take to survive the long journey
from deep time to present day?

If mere abundance was a critical factor in the survival of ancient proteins, the keratins would have a great fossil record.
Claws, beaks, feathers, hair, hoofs, nails, and scales are all made
of the fibrous protein keratin. It is a major structural component of several parts of our own body and often occurs in concentrated form. Hair is 95 percent keratin; feathers are about
90 percent keratin. It is not surprising, then, that the first report
of the purification of an ancient protein (in 1970) was that of
keratin from the hair of a 33,000-year-old mammoth preserved
in Alaskan permafrost.[19] No attempt was made to sequence the

mammoth keratin. Identification was made simply because there were no viable alternatives.

Molecular clock studies of modern keratins demonstrate that, like collagens, they evolved hundreds of millions of years ago.[20] Scale keratin is evolutionarily older than claw keratin, which in turn is older than feather keratin. Although we might think that a fossil claw or hoof would be a ready source of ancient protein, deep time keratins have yet to be recovered and sequenced. The basis for their exceedingly labile nature remains unknown. Studies have shown that modern keratin proteins degrade very quickly; on the other hand, if you have ever walked in the woods and come across an old carcass, it's nothing but bones and hair. Factors other than quantity appear to dictate whether a protein will survive preservation into the distant future.

Recent work by David Jacobs and his colleagues at UCLA helps to explain one key factor in the preservation of ancient proteins. His laboratory has obtained the oldest known invertebrate protein sequences, which hail from a 130,000-year-old coral found in the Key Largo Formation in the Florida Keys. Sequences were obtained from six different ancient proteins extracted from the limestone (calcium carbonate) coral skeleton.[21] A common characteristic of all six proteins was the acidic (negatively charged) nature of their sequences. In one of the proteins, nearly 50 percent of all its amino acids were acidic. The other proteins all had repeated sequences where an acidic residue appeared every six or seven amino acids. In living corals, the most acidic of these proteins is known to participate in the process of skeleton building or biomineralization—it binds calcium and precipitates calcium carbonate from seawater. Based on this evidence, Jacobs has proposed that proteins that can bind very tightly to the positively charged calcium in a biomineral matrix, whether it is bone, enamel, or coral, have a much greater chance of becoming an ancient protein. And what better candidates than the proteins that help form the skeleton; these proteins function inside mineralizing bone (or coral) and, as a result, become trapped inside the skeleton, where they have safe passage for thousands, if not millions of years.

A similar process appears to have preserved the oldest known ancient protein sequence, which was extracted from an ostrich eggshell collected at the Laetoli site in Tanzania—a site famous for its preservation of human footprints and just a few miles away from the famous Olduvai Gorge. In the modern ostrich (*Struthio camelus*), proteins that participate in formation of the shell, which is made of calcium carbonate or calcite, are aptly named struthiocalcins. These proteins contain a sequence of five consecutive negatively charged amino acids that form a calcium-binding domain, which allows it to tightly adhere to the shell's positively charged calcium and avoid degradation. The ancient sequence differed slightly from the struthiocalcin of the extant ostrich, but it too had a stretch of consecutive negatively charged residues, and it appears to have survived for roughly 3.8 million years.[22]

As the case of *Gigantopithecus blacki* demonstrates, tooth enamel also appears to be a favored environment for ancient proteins. *Gigantopithecus blacki* lived in the tropics, and its fossil teeth were never exposed to the freezing temperatures that are often credited for the preservation of ancient proteins. For years, paleobiologists have been frustrated with their inability to isolate proteins from fossil bones found near the equator. High temperatures were assumed to be the bad actor; proteins were too quickly degraded. But as Welker and Cappellini and their colleagues have demonstrated, proteins in enamel—the hardest tissue in a vertebrate's body, much harder than bone—are capable of being preserved for at least 2 million years. This suggests, in general, that paleobiologists should focus their hunt for proteins in the hardest tissues that they can find.

There are, however, always exceptions to the rule. One might think that, of all the tissues in a vertebrate body, blood would be the least likely to be preserved. So too, its proteins. However, this appears not to be the case. Of the 126 proteins found in the Siberian mammoth bone proteome, 26 were blood proteins. Of particular interest were the numerous proteins related to blood clot formation. While blood in a deceased body may remain liquid for days, with time, blood clots will form. Like the mammoth proteome, the proteomes of the 670,000-year-old horse and the 120,000-year-old bison also contained unusually high numbers of blood clot proteins. It is tempting to speculate that the cross-linked and aggregated proteins in blood clots are more resistant to degradation and, like ancient proteins in bone, are preferentially preserved.

Of course, it helps if the "fossil" is very young. Take, for example, Ötzi, a modern *Homo sapiens sapiens* who lived at the very beginning of the Bronze Age and died near what is now the Italian-Austrian border.[23] He was small, perhaps 135 pounds, 5 feet, 5 inches tall, with brown eyes, and lived to be about 45 years old. In the last days of his life, in the early spring, Ötzi was traversing a pass through the Alps near the top of a long two-mile-high ridge. The tip of an assailant's arrow pierced his shoulder blade. The arrowhead sliced into a major artery, which caused uncontrolled bleeding; Ötzi was dead within minutes. His killer(s) watched him bleed out, retrieved the shaft of the arrow, and left him where he died. Ötzi's body froze and was quickly covered with snow for the next 5,000 years.

Scans of Ötzi's skull revealed that his brain was largely intact.[24] Despite this, it was thought that, in addition to the arrow wound, Ötzi had incurred a skull injury just prior to death. The proteome of Ötzi's brain consisted of sequences from 502 different proteins.[25] Approximately 20 percent of the proteins were characteristic of brain tissue, such as proteins from neurons and synapses. Given the diagnosis of an injury to the skull, researchers looked for proteins that could confirm or reject their diagnosis. Among the 502 proteins, 138 had functions related to stress, wounding, and wound healing; evidence that Ötzi had indeed sustained a head injury. Perhaps the coldest case ever solved, with the help of ancient proteins.

Taxonomic identifications, determinations of sex, phylogenies, tooth development, and now a medical diagnosis: all based on ancient proteins. The more fossils we find that contain preserved proteins, the more we will find out about the history of life on Earth. The field of ancient proteomics is exploding as more scientists enter the field every year, and they continue to recover protein sequences from ever further back into deep time. Some scientists believe that one day, we'll even recover protein sequences from dinosaurs. In fact, as we will soon see, some scientists believe we already have.

8

DINO BONES

I first met Mary Schweitzer in September 2018 at the Calvert Marine Museum in Solomons, Maryland, where she had been invited to give a presentation. Tall, quiet, and self-assured, she did something that I have rarely observed in such settings. As she waited for the event to start, she moved slowly through the audience and engaged children in conversation. "Are you interested in dinosaurs?" I imagined her saying. The kids, awestruck and intimidated at first, were soon engaged in animated conversations with the famous paleontologist. She had come a long way. In 1990, she was a married fundamentalist Christian and a mother of three children. Having made the decision to pursue an undergraduate degree at Montana State University in Bozeman, she one day found herself attending a lecture by Jack Horner, the world-famous paleontologist. At the end of the class, after the other students left the room, she had the chutzpah to introduce herself to Horner as a creationist. Unfazed, and ever the contrarian, Horner invited her to audit the rest of his course despite her in-your-face confidence that she knew more about the origin of life than he did.

Her conversion to paleontology would be a costly one: she lost most of her friends, as well as her marriage.[1] But it would eventually result in her becoming a world-famous paleontologist in her own right. Soon after she completed his course, Horner suggested that she volunteer to work in his laboratory, and he assigned her a project related to his favorite topic, the growth and maturation of dinosaurs. Making very thin slices of *Tyrannosaurus rex* bones, she observed what looked suspiciously like red blood cells. Improbable. Impossible! She asked Horner to take a look. "Prove to me that they're not," he said. Every student should have professors that provide such powerfully motivating instruction.

Schweitzer eventually entered a Ph.D. program in Horner's laboratory. After receiving her doctorate in 1995, she commenced to publish increasingly astonishing, and controversial, observations. Her laboratory reported that they were able to find DNA in bone cells from a 78-million-year-old *Tyrannosaurus rex*.[2] According to their report, the bone of the dinosaur also contained amino acids, the subunits of protein, that were characteristic of collagen. She also found evidence for the blood protein hemoglobin and its oxygen-carrying moiety heme.[3] She later reported the presence of the protein keratin in the therapod *Shuvuuia deserti* and a Cretaceous bird, *Rahonavis ostromi*.[4] These initial observations were controversial, as are many of the observations that Schweitzer and her colleagues continue to make. Note, however, that controversial doesn't necessarily mean wrong. As we describe her career and her science, we should abstain from rendering a final verdict about her research. It is too early to tell whether Schweitzer's work is laying the foundation of a new and legitimate field of paleobiology or will be shown to be invalid, a dream based on inaccurate or misinterpreted data. There is, however, much to be learned about ancient biomolecules from the debate that has arisen around her work.

In 2004, Schweitzer became an assistant professor at North Carolina State University in Raleigh, where she continued her work on the characterization of improbable "soft tissue" structures from dinosaur bones. One of her most fascinating discoveries—indeed, one of the most fascinating discoveries in paleobiology in the past several decades—concerned dinosaur blood vessels. The smallest blood vessels, called capillaries, are made of a single thin layer of cells (endothelial cells) backed by a much thinner layer of supportive membrane. They can be nearly as small as the diameter of a single red blood cell, and their function is more that of selective permeability than the transport of blood. While most of our own blood vessels are found in soft tissue, such as the brain and liver, bones also contain blood vessels. Lots of blood vessels. And this is precisely where Schweitzer was able to find dinosaur blood vessels. Despite the ravages of fossilization, these structures look quite like normal blood vessels. Most importantly, they are composed of organic material—they have not been permineralized. The blood vessels are often branched, and they are flexible; they can be stretched, and, like a rubber band, they bounce back to their original length. A year after her arrival at North Carolina State, Schweitzer published a paper that described the isolation of blood vessels from the bones of a 68-million-year-old *Tyrannosaurus rex*.[5]

The process of vessel isolation is surprisingly simple now that Schweitzer has shown the way. Tiny pieces of bone are placed in a calcium-sequestering agent that tightly binds calcium; after several weeks, the mineral portion of the bone disappears, and brownish-colored sheets of organic material remain. Through this process, Schweitzer was not only able to identify flexible three-dimensional blood vessels, but also bone cells and osteocytes. Jasmina Wiemann, working with Derek Briggs at Yale, is

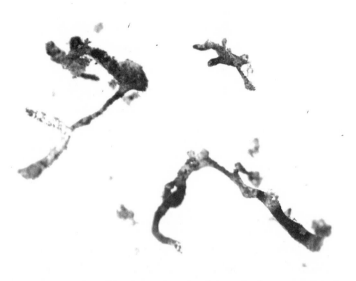

FIGURE 8.1. Ancient blood vessels isolated from the humerus of the large 225-million-year-old predatory reptile *Nothosaurus*, after removal of the bone's calcium mineral matrix.

one of a number of scientists who have been able to reproduce Schweitzer's work. In Wiemann's case, however, the ancient blood vessels were isolated from a fish and a sauropod, both Jurassic in age.[6]

While the results of Schweitzer's work with thin slices of dinosaur bones has proved revolutionary, scientists have been examining thin slices of dinosaur bones since 1850. From the mid-1950s on, nearly a hundred such reports have been published, including over a dozen by Roman Pawlicki of the Jagiellonian University in Kraków, Poland. Pawlicki described blood vessels, bone cells, and red blood cells—in which he detected high levels of iron—from an 80-million-year-old specimen of *Tarbosaurus bataar* from the Gobi Desert in Mongolia.[7] He even reported the presence of DNA, collagen fibrils, lipids, and carbohydrates.[8] The bone cells he described were osteocytes, a term applied to the bone-secreting osteoblasts once they entomb themselves in the newly made mineral matrix of the bone. Characterized by numerous thin pseudopodia, these cells have also been isolated and described by Schweitzer. Although published in prominent scientific journals, Pawlicki's work failed to achieve much publicity. It was Schweitzer's ability to isolate these cellular structures, and her subsequent characterization of their ancient biomolecules, that caught everyone's attention.

A Dinosaur Protein Sequence

In 2007, Schweitzer's laboratory reported their first ancient protein sequence. She had recovered a number of different short sequences of collagen—the longest was 18 residues long—from the femur of a 68-million-year-old *Tyrannosaurus rex*.[9] Subsequent comparison of the sequences with those from extant animals suggested that this famous and fearsome dinosaur was most closely related to birds. Criticism rolled in immediately. Scientists claimed that the preservation of intact proteins into deep time was thermodynamically impossible; that one of the sequences was incorrect; and that one of the sequences was more closely related to amphibians. Others proposed that the dinosaur collagen was a mere contaminant, perhaps from a lab technician's skin cream. After all, both collagen and keratin proteins are ingredients of popular skin creams, and Schweitzer had also recovered a keratin protein sequence from the dinosaur bone. Perhaps the source of contamination was dandruff or accidentally expirated saliva from a technician who had had eaten a chicken sandwich for lunch. Were the criticisms valid? Had Schweitzer made mistakes, or had she successfully sequenced a real dinosaur protein?

Some of her scientific colleagues called for improvements to Schweitzer's methodology. By 2007, ancient DNA researchers had sorted out many of the issues that had led to the erroneous identification of DNA in insects entombed in amber in the early 1990s. Protocols had been put in place to reduce contamination, to physically separate ancient DNA and modern DNA isolations, and to perform duplicate sequencing in different laboratories. Similar improvements, her colleagues suggested, should be applied to ancient protein research. Schweitzer initially resisted these suggestions, but she quickly adopted the new protocols and used them in her laboratory.[10] She was literally inventing the new science of deep time ancient proteomics and learning as she went along.

Over the next decade, Schweitzer's laboratory went on to report the sequences of more than a half dozen proteins from deep time fossils, mostly dinosaurs. Invariably, however, the sequences were short: tiny fragments of the ancient proteins. The most complete sequence covered 24 percent of a collagen protein from the hadrosaur (duck-billed) *Brachylophosaurus canadensis*.[11] Nevertheless, the ability to isolate proteins from isolated blood vessels and osteocytes represented a significant advance: it was much easier to recover noncollagen protein sequences in the absence of the quantitatively dominant collagens. And indeed, several of these, such as the actin and myosin, were proteins that one would expect to find in the walls of blood vessels.

Gradually, Schweitzer's professional isolation eased as other laboratories, albeit relatively few, began to report ancient protein sequences. For instance, Peter Lindgren and his colleagues reported intact collagen fibrils and the protein itself in the humerus of a 70-million-year-old mosasaur. They also reported a half dozen other proteins, including hemoglobin, collagen, and keratin, in the preserved skin of a 175-million-year-old dolphin-shaped ichthyosaur.[12] Additionally, Graham Embery and his group at the University of Liverpool reported the presence of osteocalcin in the rib bone of a 120-million-year-old Iguanodon, from which they were able to obtain a 14-residue sequence.[13] Unfortunately, they were unable to match the sequence to any known organism. Except for this single sequence, none of the proteins Lindgren and Embery identified were based on sequence data. In its place, they used antibodies to react with and identify the ancient proteins.

Despite these reports of deep time proteins by laboratories other than Schweitzer's, criticism of her work and the concept of deep time protein preservation continued. In reference to Schweitzer's work, Matthew Collins, a recognized authority in proteomics at the University of Copenhagen, has stated, "It's great work. I just can't replicate it."[14] Nor can most other laboratories. More than one of Schweitzer's scientific detractors didn't even believe that the blood vessels and osteocytes were real; they were convinced that the structures were the products of bacteria.[15]

To understand the basis of this alternate explanation, consider *Mycobacterium tuberculosis*, the causative agent of tuberculosis. This bacterium is tough to kill—current drug regimens require a combination of several different drugs for six to nine months, and if the bacterium is drug-resistant, treatment may require two years. The basis for the bacteria's resistance is, in part, what is called a biofilm. Instead of existing as individual organisms, the bacteria secrete a thick film, a protective scaffold, in which they reside. They essentially become a giant two-dimensional communal organism composed of millions of their kind surrounded and held together by an organic matrix. Ninety percent of the film is made of secreted carbohydrates, carbohydrate-binding proteins, and even molecules of

DNA; only about 10 percent of the film's mass is made of bacteria. The films, which are very persistent, can form on all sorts of surfaces in hospitals, as well as within a patient's lungs. Could biofilms, by forming on the surfaces of canals in bone that once housed blood vessels, form a structure that mimics a vessel? Were Schweitzer's ancient vessels simply fossilized bacterial biofilms? Fortunately, these were testable hypotheses.

Schweitzer and her colleagues quickly responded to the criticism of their work. Using infrared spectroscopy, Johan Lindgren identified distinct differences between bacterial biofilms and the soft tissues derived from fossil bone.[16] Schweitzer's laboratory incubated blood vessels isolated from *Tyrannosaurus* bone with antiserum to bacterial biofilms and found that it did not react with the vessels. Schweitzer then did an experiment that provided an even stronger refutation of her critics' objections. She procured pieces of modern cow bone and treated them with bleach and other reagents to remove all blood vessels and other organic material. She then used the stripped bone and its now empty canals as a scaffold on which to grow bacteria. After two weeks, Schweitzer's lab demineralized the pieces of bone and examined the material that remained behind. The bacteria, having adhered to and grown on the surfaces of the many empty canals in the bone, produced biofilms. But their morphologies were very different and easily distinguished from vessels isolated from *Brachylophosaurus canadensis*.[17] It would be convenient to conclude that the case is closed, that her dinosaur blood vessels are in fact dinosaur vessels. However, the existing evidence is far from definitive—a two-week-old cow bone is not a 68-million-year-old dinosaur fossil—and critiques of her work continue.[18]

Schweitzer's science also drew criticism from another group— the Institute for Creation Research, headquartered in San Diego,

California. The long-accepted age of *Tyrannosaurus rex* is, they claim, off by about 68 million years. Mark Armitage, a scientist at the institute, published work that demonstrated the presence of ancient osteocytes in the horn tissue of a partial *Triceratops* fossil that he collected from the Hell Creek Formation in Montana.[19] Armitage's Dinosaur Soft Tissue Research Institute then went on to reveal the presence of blood vessels and red blood cells from the specimen and, from a specimen of a juvenile *Tyrannosaurus rex*, osteocytes, blood vessels, nerve fibers, proteins, and even bacteria (unpublished).[20] Reportedly coeval with the dinosaur, the bacteria were alive and motile. Schweitzer, for whom it must be a bit ironic given her background, remains apostate.

A Basis for Skepticism

One might think that after nearly 15 years of dinosaur protein sequences, the scientific consensus would have shifted; it hasn't. Protein sequences from really deep time may occasionally make it into the pages of the *New York Times*, but they do not appear in textbooks. Paleobiologists have drawn a line in the sediment— ancient proteins go back perhaps 4 million years. There is a compelling basis for their skepticism. Derek Briggs, who has spent much of his career studying ancient biomolecules and the processes by which they are preserved, has had students who have been able to isolate blood vessels from dinosaur bones. They have not, however, been able to find proteins or fragments of proteins in those vessels. In fact, for Briggs and his colleagues, it has been quite the opposite. Briggs's graduate student Jasmina Wiemann has shown that the proteins in dinosaur bones have been cleaved, cross-linked, and otherwise derivatized to the point where sequenceable protein no longer exists.

To understand the basis of Briggs and Wiemann's work, we need to explore the reaction of sugars with proteins. One important place where such reactions occur is within our blood. The glucose in our blood slowly but constantly reacts with hemoglobin, the major protein in blood, to form what is called "glycated" hemoglobin, or A1C. If you watch much television, you have probably seen commercials that promote drugs for diabetes. "Lower your A1C," they exclaim. The higher our blood glucose levels, the higher our level of glycated hemoglobin. Whereas a clinical laboratory's test for glucose measures how much blood glucose is present at the time the blood sample was drawn, a test for A1C provides a measure of blood glucose levels over the previous several months of a patient's life and is a better test for chronic high blood sugar. A1C levels of 5 to 6 percent are considered to be normal, 7 percent is an indication of diabetes, and a value above 8 percent is diagnostic of uncontrolled diabetes.

First discovered in 1912 by the French chemist Louis-Camille Maillard, the reactions of sugars with proteins have been known as the Maillard reaction ever since. But it was the African-American chemist John Hodge who revealed the complexity of the reactions between sugars and proteins and, in so doing, created a new and important field of chemistry.[21] Pretty much everything we cook contains both sugars and proteins—and the products of their reactions with one another. The dark brown color of toast and grilled steaks is due to chemical reactions between sugars and proteins. Coffee beans, prior to the roasting that enhances sugar-protein reactions, are the color of raw peanuts. And, in addition to colors, the chemistry elucidated by Hodge gives rise to the flavors and aromas characteristic of cooked foods. In our blood, the reaction occurs at the relatively low temperature of 37 degrees centigrade. But the reaction occurs at much higher rates at elevated temperatures,[*,22] and this is where we get back to ancient proteins.

Jasmina Wiemann attempted to isolate soft tissues from the bones of 15 different deep time fossils, many of which were dinosaurs, that ranged in age from 205 to 25 million years old.[23] Not every specimen yielded soft tissue. But from those that did, the blood vessels, osteocytes, and other soft tissues were invariably stained brown, a characteristic of tissues that have undergone Maillard-like reactions. Most importantly, the researchers found that the normally linear primary structure of proteins had been transformed into three-dimensional sheet-like arrays of highly modified polymeric organic material, as a result, in part, of Maillard-like reactions. Individual amino acid subunits, let alone proteins, were absent. They had been so highly derivatized that they were no longer recognizable and certainly not sequenceable. A critical consequence of this transformation was the inability of bacteria and fungi to digest the transformed material, a circumstance that may have contributed to its preservation.

Wiemann went on to perform what she termed "experimental maturations," in which fresh chicken bones were heated in an attempt to mimic aspects of the fossilization process. Surprisingly, she found that simply heating the tissue at 60 degrees centigrade (140 degrees Fahrenheit) for as little as 10 minutes resulted in Maillard reaction–induced transformation of bone proteins and the production of what she has termed protein

---

* In 2002, the Swedish chemist Margareta Törnqvist showed that French fries and potato chips contain high levels of a carcinogen generated as a result of the Maillard reaction. When food is cooked at 200 degrees centigrade (400 degrees Fahrenheit), ten times as much of the carcinogen is formed than when cooked at 100 degrees centigrade.

fossilization products. While the exact conditions of fossilization are impossible to know—and highly variable across specimens—Wiemann's work suggests that it is simply not possible to isolate sequenceable proteins from the highly cross-linked and derivatized protein fossilization products formed by the Maillard reactions. The contrast between the data from Schweitzer and Briggs's laboratories could not be more stark.

The advances in our understanding of protein "fossilization" that came from Briggs's laboratory were quickly applied to other unsolved scientific questions. For example, it helped explain the evolution of bird eggs. Therapod dinosaurs, including those that live today, laid eggs. In fact, many different groups of dinosaurs produced eggs, as exemplified by the work of Jack Horner and his colleagues at the Egg Mountain site near Choteau, Montana. The eggs had been preserved because they were hard, calcified—similar to the eggs that we break open when we make a breakfast omelet. Eggshells, like vertebrate bones and clam and snail shells, are mineralized tissues that preserve well into deep time. Moreover, the thick crystalline calcium carbonate eggshells, like shells and bones, contain small amounts of numerous different proteins. However, paleobiologists have noticed that almost all dinosaur eggs are, in a geological sense, quite young. Where are the older Triassic hard-shelled dinosaur eggs? Contrary to the old adage, in this case, the absence of evidence may, in fact, be evidence of absence.

Of course, not all eggs are hard. I remember turning over a large pile of compost at my home in Maryland on a particularly cold spring day. At one point, the pitchfork came out with an egg that had been pierced by one of the pitchfork's tines; a large eastern rat snake quickly exited the pile of dirt and disappeared. The snake egg was whitish in color and flexible: leathery and soft. Additional uncovered eggs appeared to be partially deflated, as if a vacuum pump had been attached to them, and they had imploded. Instead of being made of crystalline calcite within which are small amounts of protein, soft eggs, like those of most lizards and snakes and many turtles, are composed of a protein matrix within which are suspended occasional crystals of calcium carbonate. One can understand why soft eggs, like most soft tissues, do not survive the journey from deep time. So, what is the evolutionary origin of hard, calcified eggs? To answer this question, Jasmina Wiemann has built on her work on ancient proteins and protein fossilization products (PFPs).

Wiemann, who is now at the California Institute of Technology, has collaborated with Mark Norell at the American Museum of Natural History and other colleagues in a unique study of ancient dinosaur egg biomolecules.[24] Using Raman spectroscopy, they examined eggshells, both soft and hard, from dozens of living organisms, along with a variety of fossil eggs. Among the fossils were the eggs of an approximately 78-million-year-old *Protoceratops* from Mongolia and *Mussaurus patagonicus* recovered from a 200-million-year-old site in Argentina. In both cases, the researchers could see obvious embryonic skeletons within the eggs, but it wasn't until they observed what they termed "diffuse halos" around the embryos that they realized they were dealing with fossils of soft eggs. In fact, when examined more closely, some of the *Protoceratops* eggs were found to be pliable!

The soft-shelled eggs of the *Protoceratops* and *Mussaurus* specimens were found to be primarily organic in nature as demonstrated by the presence of large amounts of protein fossilization products. The researchers also found that the PFPs of hard-shelled eggs were less degraded than those from soft-shelled eggs, an observation that is crucial to our understanding of the mechanism by which sequenceable proteins are preserved in younger fossils. Again, it appears that the calcium carbonate of the hard biomineralized shells, like the bones of vertebrate fossils, somehow protect and preserve an organism's biomolecules. Additional analyses determined that there seemed to be three distinct compositions, which indicated that hard eggshells had independently evolved among the dinosaurs on three different occasions.

Our ability to identify soft-shelled eggs based on unique PFPs will allow for new insights into dinosaur behavior. Living organisms that produce soft-shelled eggs, like the eastern rat snake, bury their clutch of eggs in something akin to a compost pile that, as it degrades, generates heat. In an open nest full of hard-shelled eggs, the heat required for embryonic development is provided by a brooding parent. The transition from soft-shelled to hard-shelled eggs required a major change in parental behavior. No longer could dinosaurs simply bury their eggs and wander off and leave incubation of their embryonic young to rotting vegetation.

In addition to the objections offered by Briggs and Wiemann, other scientists have argued against the deep time preservation of proteins, albeit on different grounds. For example, Jakob Vinther, Evan Saitta, and their colleagues have published important studies that question the existence of the protein keratin, the major component of feathers, hair, and scales, in deep time.[25] While it is impossible to replicate the fossilization process—humans only live so long—Vinther's team attempted to mimic deep time fossilization by treating feathers from living birds, horsehair, and scales from birds and a crocodile in two very different ways.

First, they incubated the feathers with a collection of microbes normally found on feathers for 50 days at slightly elevated temperatures. Second, they treated feathers, hair, and scales at high pressure (about 250 atmospheres) for 24 hours at temperatures as high as 250 degrees centigrade. In both cases, the results were not pretty. The feathers turned into a thick, stinky liquid that, upon analysis, contained no hint of keratin. Given these results, Vinther's group concluded that keratin had little potential for deep time preservation. While similar data from a half-dozen other laboratories have been reported, these results are rarely reported in the popular press—instead, it is those studies that yield positive results, the purported detection of keratin, that garner attention.

Many such reports are based on a technique called immunohistochemistry. In this approach, an antibody, supposedly specific for the protein that a researcher wants to detect, is incubated with fossil tissue. It binds to its target protein, if that protein is there, and the antibody, which is tagged with a radio collar-like chemical, is then visualized under a microscope. If the antibody is present, we assume its target, the ancient protein, is also present. Immunohistochemistry is a hugely valuable technique; it is used in medicine and nearly all branches of biology to detect and localize individual proteins in situ. Pathologists commonly use the technique when they need to know if a particular protein is present in biopsied tissue—for example, HER2, the presence of which indicates a poor prognosis for patients with breast cancer. Within the field of paleobiology, immunohistochemistry has been used for decades to detect ancient proteins, and many of the more recent reports of deep time proteins have relied on the technique.

One of the more interesting studies that used antibodies instead of protein sequence data to detect an ancient protein has provided insight on the origin of dinosaur flight. A bird's ability to fly depends on wings with an airfoil shape, asymmetric and overlapping feathers, and, at the molecular level, a unique type of keratin. All vertebrates make one or more of the several different types of keratin, but the keratins in modern bird feathers are particularly rich in sulfur, a component that allows keratin molecules to be linked together to form a stronger and tougher feather. Molecular divergence studies have suggested that modern feather-specific keratins evolved about 143 million years ago.[26]

Zhonghe Zhou and his collaborators, including Mary Schweitzer, wanted to see if they could detect differences between the ancient keratins in very old and much younger dinosaur fossils. They were well positioned to undertake the study because they had access to numerous fossils of feathered theropods, including species both older and younger than 143 million years. Although the scientists made no attempt to obtain keratin protein sequence data from the fossil feathers, they took advantage of their ability to detect different types of keratin using both antibodies and measurements of sulfur content. Their results demonstrated that the feathers of a 160-million-year-old dinosaur of the genus *Anchiornis* had distinctly different keratins than those found in a 130-million-year-old theropod dinosaur in the genus *Eoconfuciusornis*, which is more closely related to the direct ancestor of birds.[27] The researchers' results appeared to confirm the theory that the feathers of older dinosaurs had less sulfur, fewer cross-links, different keratin proteins, and, as a result, were less capable—or incapable—of supporting flight. If confirmed by other laboratories, this data would be an extraordinary example of the potential of ancient biomolecules to inform us about the evolution of complex behaviors.

However, Jakob Vinther, Evan Saitta, and their colleagues, aware of reports by Schweitzer and her colleagues of keratin in the feathers of another dinosaur, the bird-like therapod *Shuvuuia deserti*, re-examined material from the same specimen that had been reported to contain keratin. Using several different spectroscopic techniques, they determined that the feather-like material in *Shuvuuia deserti* was composed of calcium phosphate, not organic material, let alone keratin. Immunohistochemistry, they concluded, is simply not an accurate method by which to analyze deep time specimens.

One complication commonly encountered when attempting to identify proteins from deep time is particularly insidious. We have long known that fossils are subjected to high pressure and heat when buried and, in some cases, they are saturated with water, like those at the Fossil Bowl in Idaho. It turns out that the individual amino acid constituents of ancient proteins can react with chemicals in the water or with chemicals produced within the fossil in a way that changes the structures of the amino acids. Neutral amino acids can become acidic, positively charged amino acids can become negatively charged, amino acids with an alcohol group can have that grouped cleaved off—the list of potential modifications is long. Imagine then an antibody that is made to recognize the following sequence of four amino acids: (1) neutral, (2) positively charged, (3) neutral, and (4) an alcohol. Now imagine that the protein is transformed so that its sequence reads (1) negatively charged, (2) negatively charged, (3) negatively charged, and (4) an amino acid that is no longer an alcohol. Is there any reason to think that the antibody will bind to the modified protein? The answer, of course, is no.

The current impasse between Schweitzer and her detractors will require time, and the input of numerous laboratories, to resolve. In fact, Wiemann's work on protein fossilization products has already been criticized by scientists who ascribe her data to background luminescence.[28] The scientific community is keenly aware of past controversies in which the establishment, dogmatic as establishments tend to be, refused to acknowledge some new discovery, only to be proven wrong in the long run.

Plate tectonics is perhaps the most famous example, but my favorite is Lake Missoula, the cause of the scablands of eastern Washington State.

In the 1920s, the geologist J. Harlen Bretz theorized, correctly as it turned out, that the channels of the scablands were carved out by a series of floods that occurred about 14,000 years ago when glacial dams near Missoula, Montana, suddenly broke and released floodwaters that flowed all the way to the Pacific Ocean.[29] The geological community at the time had just spent a century convincing the world that geological changes happen extremely slowly, over thousands, even millions of years. Now, in their thinking, Bretz wanted to resurrect the idea of a Genesis-like flood. Unanimous in their beliefs, they staunchly resisted his conclusions. In the end however, Betz was proven correct. Perhaps 65-million-year-old dinosaur protein sequences will be another such story.

Recently, Schweitzer has upped the ante. Her most recent work takes aim at the deep time preservation of dinosaur DNA. During adolescent growth, cells called chondrocytes, which make cartilage, are either replaced by, or become, osteoblasts. The later cell then replaces existing cartilage to form new bone tissue. Schweitzer was well-placed to access adolescent dinosaurs. In the 1980s, Jack Horner had discovered a nesting site of duck-billed dinosaurs (*Hypacrosaurus stebingeri*) in northwestern Montana. Individual nests contained dozens of fossil nestling skeletons. When Schweitzer examined thin sections of nestling bones with a microscope, she found wonderfully preserved juvenile soft tissue that contained cartilage instead of bone. She saw what looked like chondrocytes that were in the process of dividing. She saw nuclei, and within the nuclei, chromosomes. Using fluorescent reagents that bind to and allow imaging of DNA, Schweitzer and her colleagues reported the presence of DNA in these ancient chondrocytes.[30] In fact, the stains were specific for intact double-stranded DNA—they are unable to bind to small single-stranded fragments of DNA. Might Schweitzer someday report the sequencing of dinosaur DNA?

She would not be the first. In 1994, Scott Woodward and colleagues from Brigham Young University published a short DNA sequence obtained from 80-million-year-old bone fragments of what they assumed was a dinosaur.[31] Criticism descended like a ton of bricks. Laboratory after laboratory identified the sequences as being from human contamination. Among those who criticized Woodward's results was Mary Schweitzer.[32] One can imagine that, as a result, Schweitzer would be hesitant to publish an ancient chondrocyte DNA sequence. If Schweitzer ever does publish a dinosaur DNA sequence, it must be incontrovertible. As Carl Sagan famously said, "Extraordinary claims require extraordinary evidence."

9

ANCIENT DNA'S TENUOUS ORIGINS

DNA (deoxyribonucleic acid) is the ultimate ancient biomolecule. It carries, directly or indirectly, the information required to make every biomolecule of every multicellular organism on earth, including DNA itself. Small random changes in DNA sequences are responsible for the evolution of new species and the diversification of life from simple to complex. If we possessed the genomes—the complete sequence of every gene of an organism—of only a few of the millions of deep time lifeforms that ever lived, our understanding of the history of life on Earth would expand exponentially. But we never will, because the beautiful DNA molecule, like the proverbial rose in winter, is exceedingly fragile; it is susceptible to destruction by a wide array of chemicals and environments. Despite reports of DNA in dinosaur bones, the oldest valid DNA sequence is a little less than 2 million years old, a quick blink of deep time's eye.

But does 2 million years represent a true boundary when it comes to ancient DNA? Is it possible that we could someday realize the dream of finding entire genomes of long-extinct animals? Our enthusiasm is encouraged by the seemingly endless development of new technologies. When scientists first mapped the entire human genome in the 1990s, the task required dozens of laboratories and more than a decade of work. Today, an entire genome can be sequenced, with much greater accuracy, by one person in a single day. Given such rapid advances, the generation of a genome from a fossil tooth of a *Tyrannosaurus rex* does not seem so implausible.

The idea that we could resurrect long-extinct dinosaurs using ancient DNA—obtained, of course, from the abdomen of a blood-engorged mosquito preserved in amber—was first brought to the public imagination by Michael Crichton's *Jurassic Park*. Crichton was an extremely talented and imaginative writer, but how did he come up with such an incredible idea? In the early 1980s, several individuals had independently proposed the possibility, but none of them had published anything about the concept. One of them was George Poinar, at that time a scientist at the University of California, Berkeley. Originally trained as an entomologist, Poinar quickly gained a reputation for examining small pieces of amber and finding a surprisingly varied array of deep time life-forms. In 1982, he published a paper that showed a 40-million-year-old fly entrapped in amber with its tissues so well-preserved that cellular structures, even nuclei, could be observed.[1] Shortly thereafter, Poinar and several colleagues formed what they called the Extinct DNA Study Group. The group published annual newsletters, one of which detailed their ideas about the possibility of recovering dinosaur DNA from blood-feeding insects preserved in amber.

Crichton, a medical school graduate, stumbled across the newsletter and arranged to meet with Poinar in Berkeley. Their conversation was one-sided, with Crichton listening but not revealing the reasons for his interest in Poinar's work. Undoubtedly, it was from this conversation that Crichton obtained the information that allowed him to formulate the science portrayed

in *Jurassic Park*, including the central role played by the fossilized tree resin known as amber.

If you ever have the chance to look through a microscope at
an insect preserved in amber, you are in for a treat. The preservation of morphological detail is so great compared to compressed fossils in rock, and the supply of fossils in amber is
so vast, that many paleoentomologists restrict their research
to amber specimens, refusing even to look at a flattened fossil
in rock. The ability of amber to preserve intact and detailed
morphological features—of insects, bird legs, salamanders, an
intact lizard skull, and even an ammonite—has made it an
important focus of paleobiology. It is especially important in
my field of paleoentomology, the study of fossil insects, which
is why, in 2019, I wound up attending the 8th International
Conference on Fossil Insects, Arthropods, and Amber in the
beautiful and historic city of Santo Domingo in the Dominican
Republic.

At the end of the meeting, there were field trips to several of
the amber mines on the island. In contrast to the mines portrayed in the movie *Jurassic Park*, which involved dozens of men
working in a long horizontal mine shaft with electric lighting,
beautifully engineered timber supports, and even amber processing equipment, the mines of the Mina de Valle in the eastern
region of the Dominican Republic were somewhat less glamor-
ous. The first mine I encountered was a 100+ foot deep vertical
shaft, perhaps five feet square, with few observable structural
supports. A small engine powered a winch, positioned about 10
feet above the shaft, that was slowly pulling a young man to the
surface. He clung to the rope with his hands, his feet rested on
top of a bucket at the rope's end, and he held a flashlight in his
mouth. Shirtless, sopping wet with perspiration, and covered
in clay mud, he was not smiling.

FIGURE 9.1. A 20-million-year-old fly, *Psectrosciara fossilis*, entombed in Mexican amber.

A second mine, the La Cambre mine in the northern region of the Dominican Republic, was a short horizontal hole—the term "shaft" would be an exaggeration—dug perhaps 10 feet deep into the side of a hill. Within a few moments of our arrival, a dozen or so local miners appeared, eager to sell their amber in the absence of the middlemen to whom they are dependent for most of their earnings. I was intrigued by and purchased a baseball-sized piece of raw amber that was terribly flawed, rift through with countless cracks, and mostly opaque—if it contained an entombed insect, it was impossible to see. While the specimen was of little scientific value, it was well worth $50 to be able to show friends back home that most amber does not look like the beautiful pieces one sees in museums.

Amber is itself composed of ancient biomolecules. But we are primarily interested in the biomolecules of the species it preserves. Resins, released by trees in response to physical damage, are, to a small insect, a sticky death trap. Viscous and full of

FIGURE 9.2. A typical amber mine shaft in northern Dominican Republic—not exactly like the mine visited by *Jurassic Park*'s "bloodsucking lawyer."

numerous nasty chemicals, the insects slowly become entombed and die. Given enough time, the resins, if buried in the correct environment, polymerize to form a solid. The scientific literature often portrays amber as the perfect preservational environment: the inclusion is physically protected, water is gradually lost, the insect becomes dehydrated (mummified), and the toxic components of the resin act both as an antibiotic and a natural fixative. Poinar's idea made perfect sense—DNA, the world's sexiest biomolecule, preserved in nature's perfect preservative. In the aftermath of the 1990 publication of *Jurassic Park*, it looked for a moment as if the dream was quickly becoming a reality.

A Period of Dilettantism

In the early 1990s, deep time DNA was recovered from insects in amber that were over 120 million years old. Numerous papers were published: DNA from a termite, a weevil, a bee, and a fly—all in amber. Scientists even found DNA from a dinosaur bone.[2] Published in the most respected of scientific journals, the research garnered a huge amount of publicity and fame for the authors. And why not? Analyses of DNA from related organisms in ambers that spanned a hundred million years would allow for an unprecedented understanding of the most basic mechanisms of evolution. The first publications spurred on subsequent reports, and it appeared we were on the fast track to a new dawn in paleobiology—that is, right until it all came crashing down. Fellow scientists were unable to reproduce the original data, even when the authors provided their original DNA extract for analysis.[3]

Subsequent research has demonstrated that insects entombed in very young semifossilized amber (termed "copal"), just a few thousand years old, were already devoid of DNA.[4] Specimens of insects collected just a hundred years ago, dried and stored in museums, contained less than half the DNA present in live insects.[5] Even the tough polymer that makes up an insect's exoskeleton, chitin, known to be highly resistant to degradation, could not be found in 16-million-year-old amber from the Dominican Republic.[6] The bad news continued to pour in. The entombed tissues were, in many cases, made of exogenous materials. Instead of being organic biomolecules, they were inorganic salts like pyrite and aluminum silicates, which replicated the tissues' original structures almost perfectly.[7] And in many cases, although the fossilized insects were seemingly perfectly preserved when looking at the amber from the outside, on the inside, the fossils turned out to be hollow.

The culprit for this false optimism was a new technology called the polymerase chain reaction (PCR), or more accurately, the misuse of PCR. The unprecedented sensitivity of the process, combined with poor technique, allowed for the amplification of contaminating DNA. All it took was a single molecule of DNA in a mist, vapor, or breath wafting through the laboratory to produce an erroneous result. A DNA sequence thought to be from a 120-million-year-old weevil was eventually shown to be related to DNA from a fungus. DNA from living fruit flies, crickets, and grasshoppers, but not the insects entombed in the piece of amber, were detected by the PCR procedure. These problems were foreseeable—the issue of contaminating DNA was well known by 1990—but an unwarranted optimism led otherwise level-headed scientists to overlook the possibility. One critique aptly deemed the early 1990s a period of scientific dilettantism.[8]

In the wake of these embarrassments, conferences were held and standardized rules were instituted. To avoid contamination, it was recommended that ancient DNA be processed in a separate dedicated laboratory, and that any results be reproduced and confirmed by independent laboratories.[9] Today, the critical role of the polymerase chain reaction in the establishment of legitimate ancient DNA research is universally accepted. In fact, despite the several embarrassing burps at the beginning of ancient DNA research, PCR has been, in large part, responsible for our progress towards ever older and ever more complete ancient DNA sequences.

Indeed, it is difficult to portray the degree to which modern biology depends on, and has benefited from, the use of PCR. To understand exactly why this tool is so powerful, and why its misuse resulted in the false claims of the early 1990s, it is worth learning a bit more about how the technique works.

The PCR process, for which Kary Mullis was ultimately awarded a Nobel Prize in Chemistry, allows specific portions of our DNA to be selectively picked out from trace amounts of unruly mixtures of DNA and greatly multiplied. The seemingly infinitesimally small amount of DNA left on a cigarette butt, for example, can identify a criminal after the application of forensic techniques based on PCR. In the mid-1980s, Mullis had developed a preliminary version of the PCR process but, at that point, it was far from the game-changing—and commercially viable—technique that it would later become.[10] The process involved mixing together the DNA sample, reagents, and a small amount of a protein, the enzyme DNA polymerase. This enzyme takes small pieces of DNA and replicates them to make double-stranded DNA. After completion of the first round of the reaction, the mixture had to be heated (to 95 degrees Celsius) in order to separate the two strands of newly formed DNA so they could become templates for a second round of duplication. Calculations have indicated that after 30 such cycles of amplification, as many as a billion copies of a single target DNA sequence could be produced. Even trace amounts of DNA become easily identifiable when they have experienced a billionfold increase. (This is also why unknown contaminants can wreak such havoc.)

Mullis, however, had a serious problem that stymied the development of his invention. The process required the addition of new DNA polymerase to initiate each cycle, as heating the mixture destroyed the enzyme. The required additions greatly slowed the process and increased the number of physical manipulations needed to perform PCR. What was needed was a heat-stable enzyme, one that would not be destroyed when the temperature was increased to 95 Celsius. Fortunately, as other scientists would soon realize, the enzyme they needed had been found 20 years earlier in Yellowstone National Park.

In the mid-1960s, a bacteriologist at Indiana University, Thomas Brock, initiated a 10-year research program to study

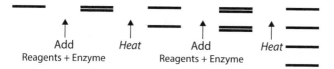

FIGURE 9.3. A flow chart that depicts a 4-fold amplification of a segment of single-stranded DNA via two cycles of PCR. In each cycle, reagents, including a DNA polymerase enzyme, replicate the DNA to form a double-stranded polymer, which is then separated by heating in preparation for the next cycle of replication.

microorganisms that live in near-boiling water in the famous hot springs of Yellowstone. At that point in time, the scientific consensus was that bacteria, and their proteins, could not survive at temperatures above 73 degrees centigrade. Operating out of a field laboratory in West Yellowstone, Brock challenged this limit and developed new techniques for the isolation and culture of what are called extremophilic bacteria. As part of this research, Brock and Hudson Freeze, an undergraduate student working in his laboratory, isolated a new species of bacteria, *Thermus aquaticus*, from material collected from the Mushroom thermal spring in the lower geyser basin of the park.[11] In subsequent work, they demonstrated that enzymes from this bacterium were able to maintain their function even in boiling water. Brock made cultures of *Thermus aquaticus* available to the public, which, nearly 20 years later, allowed the scientists responsible for developing modern PCR to solve their problem.[12] The DNA polymerase they eventually cloned is now referred to as the Taq polymerase, the acronym Taq coming from *Thermus aquaticus*. The use of the heat-stable enzyme eliminated the need to stop the PCR reaction after each cycle to add new enzyme: now we simply add the reagents to a tube and let it sit in a device that cycles through programmed steps of heating and cooling.

The Oldest Ancient DNA

The idea that ancient DNA could survive from the deep past was not itself misguided. In fact, even by 1984, there was precedent for the recovery of ancient DNA, albeit from the fairly recent past. Scientists had isolated valid ancient DNA from an Egyptian mummy, as well as a museum specimen of an extinct subspecies of the plains zebra.[13] Still, as recently as 20 years ago, the technical limit on DNA retrieval was thought to be about 40,000 years, right around the time of the disappearance of Neandertals. Fortunately, that estimate turned out to be overly pessimistic. But to see just how far back ancient DNA can take us, we'll need to visit one of the coldest places on Earth.

The peripheral waters of the Arctic Ocean are composed of eight different seas—the Norwegian, Barents, Kara, Laptev, East Siberian, Chukchi, Beaufort, and McKinley Seas—each demarked by an island or mainland peninsula. Between the Laptev and East Siberian, at 75.3 degrees north, lie the New Siberian Islands, where permanently frozen ground, or permafrost, is common. The earth's existing reservoirs of permafrost constitute a treasure trove of Pleistocene fossils, from which we have recovered the DNA of wolves, camels, mastodons, sloths, cave bears, bison, and many other animals. Add to this list mammoths from Siberia, with tusks so well preserved that their quality is equivalent to that of extant ivory. In one particularly striking photograph of the region, taken by Evgenia Arbugaeva on Bolshoy Island, a mammoth skull rests on a beach, its single 15-foot-long tusk extending skywards from the skull in a beautifully sculpted arch. In the background, framed between the mammoth's skull and tusk, a modern-day tusk hunter approaches, a small protesting wave cresting at his feet. It's an image that connotes how closely connected we are to our distant past.

The genus *Mammuthus* is thought to have evolved from an African ancestor about 5 million years ago. In northern Europe, they then radiated into several species, one of which, the steppe mammoth (*Mammuthus trogontherii*), appeared about 2 million years ago. Studies based on morphological analyses suggested that *M. trogontherii*, which inhabited both the New and Old Worlds, gave rise to the Columbian mammoth (*M. columbi*) about 1.5 million years ago in North America and the mammoth with which we are most familiar, the woolly mammoth (*M. primigenius*), about 700,000 years ago in Eurasia.[14] For nearly a half million years, the three species comingled in North America. That was, at least, the theory.

In a recent groundbreaking study by Tom van der Valk and his colleagues at the Centre for Palaeogenetics in Stockholm, several mammoth teeth were collected from northeastern Siberia, the oldest of which was estimated to be about 1.65 million years old.[15] The ancient DNA that was isolated and sequenced from these specimens demonstrates that the ancestor of Columbian mammoths was, in fact, not *Mammuthus trogontherii*. In addition, *Mammuthus columbi* appeared about 420,000 years ago, much more recently than the 1.5 million years previously assumed. The data generated by van der Valk and his colleagues suggest that *M. columbi* originated in North America, as a hybrid of the woolly mammoth and a previously unknown, and still unnamed, species.

But how could that be—aren't all hybrids sterile? The answer is complicated. While the familiar hybrid of a horse and a donkey, the mule, is almost always sterile, the well-known coywolf of northeastern North America, which originated more than a hundred years ago, appears to be a stable morphotype. Larger and stronger than the coyote (*Canis latrans*), about 25 percent of its DNA is from the wolf (*Canis lupus*). But scientists disagree as to whether the coywolf is a distinct species, and it has not been officially designated as such. Although most hybrid animals are sterile, there is a phenomenon called hybrid speciation, in which hybrids give rise to a new species. It is more common in plants than animals (although even in plants it is still very rare), but it must have happened as the two species of mammoths lived side-by-side in North America more than four million years ago.

The most interesting aspect of the study, though, is that the mammoth DNA was more than twice as old as any ancient DNA discovered to date. The previous record holder, reported in 2003, was isolated from a fossil horse from the central Yukon region of Canada. Eske Willerslev and his colleagues were doing fieldwork at the Thistle Creek site in central Yukon Territory, a site famous for its rich and diverse fossils of Pleistocene vertebrates, when they discovered the leg bone of an extinct horse, *Equus scotti*, which had been frozen in permafrost for about 700,000 years. Subsequent DNA sequencing of the bone provided the organism's entire genome and a treasure trove of data.[16]

Access to the entire genome—more than 20,000 different genes—of a deep time animal can provide huge amounts of information, including an organism's sex, skin pigmentation, and eye color. Different variants of one single gene are responsible for different hair colors, so if the sequence of that single gene can be determined, the color of a mammal's hair can be ascertained. But to really understand the evolution of *Equus scotti*, its ancestors, and its younger relatives, the sequences of the extinct horse had to be compared to other specimens. The genomes of a modern domesticated horse (*Equus caballius*) and an extant wild (Przewalski's) horse were sequenced, as was *Equus lambei*, a 43,000-year-old fossil collected from the Taymyr Peninsula on the coast of the Arctic Ocean in Siberia. Sequence differences between these various lineages indicated that the genus *Equus* first appeared a little over 4 million years ago, that modern horses separated from Przewalski's horse about 55,000 years ago, and that they were domesticated about 5,500 years ago. And this was just a tiny fraction of what researchers were able to deduce from the horses' genomes.

In addition to documenting when a species first appeared and the identity of its direct ancestors, analyses of ancient DNA can also tell us why and how a species became extinct. Consider, for instance, what ultimately became of the Siberian mammoths. The water between the Siberian mainland and the islands is currently less than a hundred feet deep. But 20,000 years ago, due to glaciation, the ocean levels were about 400 feet lower; as a result, the islands became connected to the mainland and populated with the mainland's large vertebrates. About 800 miles to the east, Wrangel Island became home to a variety of mammals, including as many as a thousand mammoths. Approximately 9,000 years ago, however, sea levels rose, and the mammoths became island bound. Isolated, they were the last mammoths on Earth, their last generations perhaps contemporaries of the Mesopotamian king Gilgamesh and the pharaoh Khufu, builder of the Great Pyramid of Giza. Genomes obtained from preserved bones of several of the last surviving woolly mammoths on Wrangel Island demonstrated a nearly 40 percent decrease in genetic variation in genes of white blood cells, which help fight off infections and diseases.[17] Lack of diversity, usually as a result of

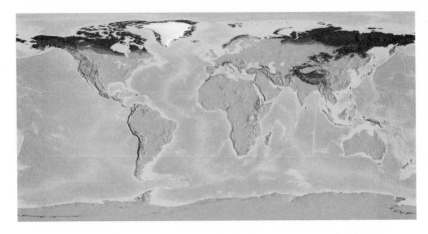

FIGURE 9.4. Earth's areas of permafrost. The shading, from lightest to darkest, represents no, sporadic, isolated, discontinuous, and continuous permafrost.

dwindling populations and subsequent inbreeding, is a slippery slope. Faced with the nearly impossible task of increasing genetic diversity, animals lose their resilience and ability to adapt, and eventually become extinct. The small and isolated population of mammoths, having survived 6,000 years longer than mainland populations, gradually disappeared. Inbreeding, while perhaps not the sole cause of their demise, played a significant role.

Frozen carcasses of much older organisms, animals that lived long before mammoths evolved, may be waiting to be found. The oldest permafrost, estimated to be 20 million years old, is located within the Transantarctic Mountains. Unfortunately, warming global temperatures—rising much faster at the earth's poles—will cause a massive thawing of the planet's permafrost. While this warming will provide greater access to numerous invaluable specimens, it also means that a large percentage of specimens currently buried in permafrost will be lost forever. Even now, exposed carcasses are rotting in the sun.

Some very good recent news suggests that specimens preserved in permafrost are not our only option. Axel Barlow, Michael Hofreiter, and their colleagues have published a high-quality genome of a cave bear, *Ursus praekudarensis*, the skull of which was found in a temperate cave.[18] Located in the Caucasus Mountains of Georgia, it was preserved far away from the permafrost of Siberia, yet survived intact for about 360,000 years. The key, in this case, was the portion of the skull from which the ancient DNA was isolated, the petrous bone. This particular part of the skull is very dense—it functions to protect the inner ear—and, as we will discuss further in the next chapter, is particularly adept at preserving its DNA. Barlow and his team were able to perform molecular clock analyses of the bear's DNA sequences and establish a phylogeny for this particular group of *Ursus*, which, despite going extinct 24,000 years ago, is a sister group to the modern brown and polar bears.

As we sequence the genomes of more and more animals, these long-extinct species will become alive in ways that we could not have imagined a few years ago. As a result, I suspect that popular infatuation with ancient life will grow far beyond the physically intimidating bulk of *Tyrannosaurus rex*.

Ancient Proteins from Deep Time DNA

Beyond merely allowing us to understand the evolution and demise of species, ancient DNA can also be used to reproduce ancient life—or, at least, one facet of it. If high-quality ancient DNA can be obtained, DNA that provides entire gene sequences, then the corresponding deep time proteins can be produced in the laboratory. Even without the entire genome, so

long as we have enough short but overlapping ancient sequences of a particular gene, the complete gene sequence can
be determined. This task is made easier if we have, as a guide,
the sequence of the gene that encodes our protein of interest
from a living organism that is closely related to the extinct animal
(e.g., for a mammoth, the elephant). But to produce the ancient
protein, we need more than merely ancient DNA. We also need
a protein factory: cultures of bacteria into which we have inserted a copy of the gene. In the early days of molecular biology,
an intact gene of interest, as "naked DNA," was simply incubated
in the presence of bacteria in the hope that some of it would be
transported into the bacterial cells and used to make proteins,
as if it were one of the bacterium's own genes. Since, then, however, we have learned that a much more efficient mechanism is
to make use of a vector, or gene carrier, to insert the ancient
gene inside the cell and mediate production of the protein.

This advance is due in large part to Joshua Lederberg, who
won the Nobel Prize in 1958 for the discovery of the ability of
bacteria to exchange genetic material with one another, a process termed bacterial conjugation.[19] The DNA that is exchanged
is in the form of a circular double-stranded DNA molecule,
which Lederberg called a plasmid. Modern molecular biologists
have co-opted this process to allow manipulation of their gene
of interest. A plasmid that is modified in the laboratory so that
it contains foreign DNA, called a vector, is manipulated so that it
contains other genes and DNA sequences that allow for the
correct processing of the foreign DNA in the bacterial cell. Create the vector, insert the vector carrying the foreign DNA (in
our case, an ancient gene) into a bacterium, and let the bacteria
synthesize the deep time protein. The manipulated DNA and
the proteins made from that DNA are referred to as recombinant
DNA and recombinant protein.

The process is similar to that used by gene therapy companies to treat genetic diseases. The biotech company bluebird bio, for example, uses manipulated viruses as vectors to treat a disease with the imposing name adrenoleukodystrophy. Patients with this disease have a mutated version of one particular gene that encodes a dysfunctional transport protein. Normally, the protein transports excess and unwanted saturated fatty acids into organelles where they are degraded. In the absence of functional fatty acid transporters, fatty acids accumulate, often in the brain, where they cause severe brain damage and, eventually, death. Bluebird bio inserts the gene for a normal fully functional transporter protein into a genetically engineered virus that then serves as their vector. Lentiviruses, which have a long history of effectively infecting humans (AIDS, for example, is caused by a lentivirus), are widely used in the field of genetic therapy, but only after the virus is genetically engineered to eliminate its pathogenic properties. Once the recombinant virus infects the patient's bone marrow stem cells, the new gene becomes a permanent part of the stem cells and their progeny and mediates production of normal, albeit recombinant, transporter proteins for the life of the patient; the patient is cured.

This technology has recently been used in a study that, perhaps more than any other work on ancient biomolecules, exemplifies the power and potential of future molecular paleobiology. Kevin Campbell, Michael Hofreiter, and Alan Cooper knew that mammoths had long hair and ears much smaller than those of modern elephants. But external morphological characteristics were surely not the only adaptations that the mammoths had evolved during their transition to the frigid environment of the northern plains of North America and Siberia. The researchers hypothesized that specific proteins had mutated as part of cold temperature adaptation. The researchers knew that when walking in snow, peripheral anatomical structures, such as the foot pad of a mammoth, were cooler than internal organs; they also knew that hemoglobin releases its oxygen less efficiently at cooler temperatures. Given these two facts, might mammoth hemoglobin have evolved to ensure adequate oxygenation of peripheral body parts? The team of scientists extracted DNA from 31 grams of bone powder obtained from the femur of a 43,000-year-old Siberian mammoth and amplified the genes for the two protein subunits that made up the mammoth hemoglobin molecule. Both genes differed from their modern elephant counterparts as a result of several mutations.

The team then transfected cultures of bacteria with the mammoth hemoglobin gene and let them make mammoth hemoglobin.[20] An ancient protein, unique to the extinct mammoth, was produced in quantities that allowed testing of its functionality. Characterization of the two ancient molecules demonstrated that one of the hemoglobin subunits had three amino acid differences relative to the subunit from Asian elephants. A substitution at position 101 was fairly radical; it involved replacement of a negatively charged amino acid with an uncharged amino acid, which caused the hemoglobin molecule to change its shape. Assays of the ancient protein demonstrated that while mammoth hemoglobin functioned normally (i.e., the same as hemoglobin from an elephant) at 37 degrees centigrade, at cold temperatures it released its oxygen load much more readily than hemoglobin from an elephant. The team had demonstrated that sequence changes in mammoth hemoglobin were part of the animal's adaptation to cold temperatures. While modern science may not (yet) be able to reproduce complete ancient organisms, it has taken the first steps in that direction: reproducing and characterizing individual functional components of long-extinct organisms.

Entire genomes have been determined from a number of ancient organisms. As a result, tens of thousands of ancient protein-encoding gene sequences are available. Chances are that many of these ancient proteins, despite a few random amino acid substitutions, function in an identical fashion to those in related extant organisms. When the complete sequence of the bone protein osteocalcin was obtained from a 42,000-year-old horse, it was found to be identical to that of modern horses. In fact, given that the osteocalcin sequences of the horse, donkey, and zebra are identical, and the horse lineage is thought to have diverged from the zebra/donkey lineage about 1.2 million years ago, no amino acid substitutions have occurred in the *Equus* osteocalcin protein for more than 1.2 million years.[21] Nevertheless, minor genetic differences in other ancient genes will have major implications. Consider, for example, what ancient DNA has recently taught us about the bubonic plague.

The first record of the bubonic plague is from an epidemic that occurred in the Eastern Roman Empire between the years 541 and 544; over 25 million people died. Molecular clock analyses have indicated that the bacterium *Yersinia pestis*, the causative agent of the plague, was derived from a relatively nonvirulent ancestor that appeared much earlier, approximately 5,800 years ago. In fact, *Yersinia pestis* underwent a number of changes before it became the virulent scourge that it is today. In a massive study, Eske Willerslev and scientists from 17 other institutions sampled the teeth of 101 Bronze Age skeletons and isolated and sequenced ancient DNA from each specimen.[22] They also isolated and sequenced ancient *Yersinia pestis* DNA from the skeletons. One of the Bronze Age bacterial genes was known to encode a protein that in today's *Yersinia pestis* affects host blood clot formation. The entire sequence of the gene was obtained from three specimens of widely separated ages: 2,966, 3,686, and 4,782 years old. In all three specimens, the ancient protein sequence differed from that of today's *Yersinia pestis* at position 259—a single amino acid residue had been replaced by another.

In the same year that Willerslev and his colleagues published their study, Wyndham Lathem and colleagues from Northwestern University in Illinois created a new strain of *Yersinia pestis* that inserted the ancient gene identified by Willerslev in place of the modern bacteria's corresponding gene.[23] They infected mice with the new version of the *Yersinia pestis* bacterium, and found that it was a hundredfold less effective at producing an invasive infection than the modern version of *Yersinia pestis* that continues to cause deaths even today. The increased ability of the modern *Yersinia pestis* to dissolve blood clots is thought to help the bacteria spread throughout the victim's body, although scientists do not fully understand why this is. What we do know is that sometime between 2,966 and 1,479 years ago, the gene that encodes this one particular *Yersinia pestis* protein underwent a single mutation that greatly enhanced the bacterium's ability to cause a much nastier version of the plague.

Given its awesome potential, ancient DNA nearly monopolizes the new science of ancient biomolecules. Ancient DNA is rapidly enhancing our understanding of the history of life on Earth, and in some cases, even allowing us to reproduce ancient molecules. As we will see in the following chapter, this includes our own evolution. But there is one nearly insurmountable obstacle that the field of ancient DNA faces: the near total absence of ancient plant DNA.

Not So Ancient Plant DNA

Nearly all life on Earth is dependent on photosynthesis. This process, mediated by the green pigment chlorophyll, involves the capture of light energy to fuel transformation of the gas carbon dioxide into sugars (for the plant) and oxygen (for us). Within the cells of a leaf, the green pigment is found within microscopic structures called chloroplasts. Though packed with chlorophyll, chloroplasts also contain DNA, and as a source of DNA, chloroplasts have a huge advantage. Most human cells have two sets of chromosomes and therefore two copies of each gene. But every cell in a leaf may have as many as 200 chloroplasts, each with a compliment of about 130 genes that have functions related to photosynthesis—and there are hundreds of copies of each chloroplast gene. A single plant cell may have 10,000 copies of a gene. If you are going to find ancient DNA, what better place to find it.

And, in the early 1990s, scientists did. The 16-million-year-old leaves from the Fossil Bowl site in Clarkia, Idaho, were incredibly well preserved. As discussed in chapter 3, some were even still green (or, more realistically, sage). And microscopic examination of the leaves revealed the preservation of chloroplasts. Given reports of isolation of DNA from insects in amber, paleobotanists were presented with an opportunity they could not resist: an alluring chance to prove that ancient DNA could be isolated from deep time plants. Numerous publications reported chloroplast DNA sequence data from, for example, magnolia and bald cypress leaves from the Clarkia site.[24] Chloroplast DNA was also extracted from 150,000-year-old peat in western Japan, and their sequences matched those from living species of fir trees in the region (*Abies*).[25]

Unfortunately, it was the amber inclusions debacle all over again. Contamination, detected and amplified by PCR, was the source of the "ancient" DNA. The reports of DNA from plant fossils were met with immediate, rigorous, and, in the end, decisive criticism. In order to compare the ancient fir tree DNA sequences with sequences from living trees, the laboratory in

Japan had extracted and sequenced chloroplast DNA from five different local living species of *Abies* in the same laboratory, at the same time, and with the same equipment as that used in processing the sample of ancient peat. No wonder ancient pollen grain sequences from the Pleistocene peat matched those of modern trees. Fifteen years later, Sascha Liepelt and his colleagues from the Philipps-University Marburg in Germany produced the first legitimate plant DNA from fossil wood.[26] The scientists wore disposable gowns, gloves, shoe covers, caps, and masks, and worked in a laboratory with positive air pressure and ultraviolet lights that destroyed contaminating DNA. Two different laboratories were used for extractions and DNA amplifications, one for fossil specimens and a separate lab for living organisms. Ten wood specimens, spanning ages of 300 to 11,500 years old, were analyzed for chloroplast DNA. Only three specimens were found to contain DNA, the oldest merely 1,000 years old.

The oldest date now widely accepted by the scientific community for ancient plant DNA is from a midden from Western Australia dated to approximately 30,000 years ago.[27] While there were no surviving remnants of leaves or sticks, the midden was assumed to have been built by a small herbivorous mammal. DNA sequences from seven different plant families were obtained. Ancient plant DNA has also been isolated from the stomach contents of Pleistocene Siberian mammals preserved in permafrost, including a horse, a bison, and a mammoth, as well as from fossilized excrement of a 14,000-year-old sloth and a 22,000-year-old mammoth.[28] Nothing though from deep time. The current consensus is that deep time plant DNA is simply not preserved. But optimism reigns supreme. When David Tank, at the University of Idaho, was introduced to the leaf fossils of Clarkia, he became infatuated with the possibilities. But Tank is a plant systematist, rather than a paleobiologist, and, to his credit, he realized that his initial attempts to retrieve ancient plant DNA, performed in a lab not specifically designed as an ancient DNA laboratory, were unlikely to be successful. Nevertheless, when I asked him about his experience, he immediately responded, "Give me $5,000 and access to an ancient DNA lab, and I would love to try this again."

So why does animal DNA extend so much further back into deep time than ancient plant DNA? One might think that the incredibly tough external coat of pollen would form a protective environment for DNA. And given the extensive fossil record of pollen, DNA from a tiny portion of that record would provide a windfall of phylogenetic information about the plant kingdom. But aside from pollen DNA from Ötzi's stomach and 6,000-year-old pine pollen from Scandinavia, the record of in situ ancient-pollen DNA is abysmal.[29] Like the deep time invertebrate organisms preserved in the Burgess Shale, plant leaves are composed of "soft tissue." It may be as simple as the fact that plants lack the DNA sequestering biominerals of bones or teeth or shells, the sources of nearly all deep time animal DNA.

Or do they? Charophyta is a group of plant-like green algae, familiar to readers who use them to create an interesting environment for fish kept in aquaria. Charophyta are, in fact, sister to all advanced plants—that is, they and plants have a direct common ancestor. They are interesting to paleobiologists because some genera of charophytes encapsulate their fertilized eggs (oospores) in geometrically beautiful structures made of calcium carbonate. The tiny shell-like structures function to protect the oospores as they lie dormant in mud waiting for an opportune time to germinate. And they have a phenomenal fossil record that goes back 425 million years.[30] Like bone, teeth, and eggshells, they contain positively charged calcium to which ancient DNA and proteins have been shown to bind.

FIGURE 9.5. The fertilized eggs (oospores), of a common type of green algae, that are encased in a calcite shell. Pictured here are fossil oospores (paleontologists call them gyrogonites) from Catalonia, Spain, that are about 70 million years old. Scale bar = 0.2 mm.

Do fossil charophytes contain ancient DNA? We don't know. In fact, we don't even know if archival museum specimens have retained intact DNA or proteins. This situation may be the result of disciplinary silos, as there is a near complete failure of botanists to talk to paleobotanists. When botanists examine the morphology of an oospore, the first thing they do is throw away the calcite shell; it is of zero interest to them. On the other hand, the paleobotonist has nothing to work with other than the shell. They even have their own paleo-specific terminology for the shell (gyrogonite), of which they have identified many different genera and species. Until those who study the past and present begin to collaborate, we will be unlikely to crack the mystery of whether gyrogonites contain ancient DNA.

Perhaps, however, the key to finding ancient plant DNA has been right under our nose this entire time, as there are now reports that ancient DNA can be obtained from dirt. Most of the soil in Greenland, as the Icelandic settlers discovered in the year 985, is buried under an icefield, 1,500 miles long and, in places, as much as two miles thick. However, tens of thousands of years ago, during some of the more protracted interglacial periods, portions of Greenland were covered by large forests of pine, spruce, and yew trees. We know this because in 2007, Eske Willerslev and an international group of collaborators were able to access samples of the silty bottoms of Greenland's ice cores. Although the cores contained nothing resembling fossils (nothing in situ), Willerslev's team was able to extract ancient DNA from the frozen mixture of ice and silt. Dated to between 450,000 and 800,000 years ago, the DNA contained sequences from trees, ground cover plants, and insects.[31] To say that this was a surprise to science is a huge understatement. But the researchers included numerous controls to eliminate the possibility of contamination; the data, sequences of which are referred to as eDNA (environmental DNA), appear to be legitimate.

In the same year that Willerslev's team found eDNA at the bottom of a two-mile-long core in Greenland, Steven LeBlanc, director of collections at Harvard's Peabody Museum, came across specimens of *Yucca* plants from the American Southwest that had been at the museum for a hundred years. Some of the specimens were 2,000-year-old-remnants of leaves that had been chewed, presumably for medicinal purposes, and then spit out. Curious if they might contain human DNA, he put together a team that went on to isolate exactly that.[32] At that time, the DNA's preservation was thought to be due to the dry conditions of the desert where the remnants of the quids were found. A good theory, but a dry environment is not necessarily a requirement for DNA preservation. More recently, eDNA has been found in the wet ocean sediments collected from a depth of three miles.[33]

Scientists have also recovered the complete genome of a 5,700-year-old Danish girl from birch bark resin that she had been chewing, like we would chew gum today,[34] as well as a partial human genome from much older, 10,000-year-old gum found in Sweden.[35] We can conclude, from these and other studies, that ancient DNA is not as fragile as we assumed. One particularly impressive study of ancient eDNA as it relates to humans comes from the Max Planck Institute for Evolutionary Anthropology in Leipzig, Germany. Researchers there found eDNA in the dirt floors of seven different caves in Europe and Central Asia.[36] Some of the caves harbored the eDNA of as many as 10 different game animals, and four of the caves contained 50,000-year-old-human DNA. In one of the caves, both Neandertal and Denisovan DNA were found. This work, along with the other ancient DNA research, has given birth to some of the most revealing science of the last century, the molecular bases of the evolution and dispersion of the human species.

10

OUR INNER NEANDERTAL

North America is, for the most part, a nation of immigrants (about 97.8 percent of us). In Japan, however, pretty much everyone's ancestors are Japanese. That doesn't mean that the Japanese are uninterested in their heritage: who wouldn't want to find out if they are descended from a famous seventeenth-century samurai? And what Brit doesn't want to know if their ancestors included a Pict, Celt, or Gael?

The advent of high-throughput sequencing technology has been a boon to modern biology. It has also given rise to a wide range of DNA-based genealogy services. With merely a credit card and some spittle, an individual can determine their ancestral heritage. A person can even determine how much Neandertal DNA they carry. Finnegan Marsh, a research and collections assistant in the Paleobiology Department at the Natural History Museum in Washington, DC, proudly displays his Neandertal pedigree on the door of his office; he is 2.9 percent Neandertal—as are most Caucasians, give or take a few tenths of a percent. While the previous chapter exposed the limits of what ancient DNA can tell us about truly deep time, here we seek to demonstrate the seemingly unlimited potential of ancient DNA to shed light on the ancestry of our species, *Homo sapiens*.

Svante Pääbo, a scientist at the Max Planck Institute for Evolutionary Anthropology in Leipzig, Germany, has spent over 30 years studying the DNA of Neandertals.[1] He and his colleagues have sequenced the entire genome of numerous Neandertal specimens. The Neandertals occupied large parts of the Palearctic, from Spain to Siberia, from about 550,000 to about 40,000 years ago. On average, they were shorter than modern humans (about 5 feet, 5 inches tall), more robust, and expended energy at a rate significantly higher than that of modern humans.[2] Pääbo's efforts allow us to ask some rather simple questions: What exactly is "Neandertal DNA"? Is it restricted to one particular chromosome? Do the husbands of the dozen or so women who contacted Pääbo to offer up their spouses for study as living Neandertals have more Neandertal DNA than we do?

Consider for a moment the diversity of *Homo sapiens* in the world today. The Han Chinese, the Yamato of Japan, the Inuit, Bengalis, Polynesians, and the numerous indigenous peoples of Australia and the Amazon are but a minuscule fraction of our species' ethnic diversity. And then consider that the genetic diversity within Africa today is greater than that of any other continent on Earth.[3] The genus *Homo*, which appeared in Africa more than 2 million years ago, has had plenty of time to diversify, though only a fraction of that diversity made it out of the continent. Although there are competing theories and disagreements about specific dates, DNA analyses of the genomes of numerous specimens suggest that there were three different out-of-Africa migrations. In the initial exodus, *Homo erectus* is thought to have arrived in Eurasia as far back as 1.85 million years ago.[4] The species was long-lived: fossils found on the Indonesian island of Java document its presence as recently as 108,000 years ago.[5]

In a second migration, approximately 700,000 years ago, an ancestral population, the Neanderovans, spread over Eurasia and eventually split into the Neandertals and the Denisovans.[6] Fossils of modern humans have been found in Africa that are about 315,000 years old, but this group didn't leave the African continent until about 60,000 years ago.[7] Genetic data show that there was interbreeding between these various groups at pretty much every opportunity. As modern humans left Africa, they encountered and genetically displaced both the Neandertals in western Eurasia and the Denisovans in the east. Exchanging genetic material as they moved south, through islands of southeast Asia, they eventually reached the island of Sahul. This huge landmass, the result of low sea levels that connected Australia to New Guinea, was populated sometime between 65,000 and 55,000 years ago.[8] While my colleague may brag about his 2.9 percent Neandertal bloodline, the Australo-Papuans put him to shame, with about 5 percent Denisovan DNA.[9] We know all of this because of the ability of ancient DNA to persist for hundreds of thousands of years.

There are innumerable levels of genetic ancestry; from our Neandertal DNA content to the prevalence of Norwegian DNA in the citizenry of northern Minnesota. Entire books have been written about the ancestry of our species based on ancient DNA sequences, including David Reich's authoritative *Who We Are and How We Got There*.[10] Rather than attempt to repeat what scientists far more knowledgeable than I have already written, my aim in this chapter will be to describe several studies of ancient DNA that document aspects of ancient *Homo sapiens* behavior and physiology. Our ability to procure such information has led to an explosion of research in the fields of archeology and anthropology, fields for which the potential of ancient DNA research is seemingly unlimited: specimens are abundant, and their bodies and genomes reasonably intact—none more so than the 5,000-year-old murder victim we previously met in chapter 7, Ötzi.

Ötzi's Last Meal

No one knows if Ötzi knew that he was in danger or even saw his murderer(s). Was he ambushed, or was he running away in a futile attempt to escape? He had 14 arrows in his quiver, but only two were armed with flint arrowheads; his bow was new and not yet operational. Whatever happened prior to the assault, the assailant's arrowhead opened a major artery that caused uncontrolled bleeding and a very quick death. His frozen and eventually mummified body provided a perfect environment for the preservation of his DNA.

Studies of the genome recovered from Ötzi's mummified body have indicated that even if he had uneventfully made it through the high mountain pass, he may not have lived much longer. He was a poster child for coronary heart disease:[11] He had several genes that are strongly associated with stroke, heart attack, and atherosclerosis—a genetic trifecta. Indeed, a CT scan of his body showed large plaques in several of his arteries, including the aorta.[12] Despite his heart problems, Ötzi possessed impressive physical prowess. How many of us have the ability to hike over a two-mile-high mountain pass? What Ötzi didn't have as he trekked through the Alps, however, was a Denisovan gene that today helps Tibetans avoid high altitude sickness, an affliction that has felled many amateur climbers attempting to summit Mt. Everest. The gene prevents the increase of red blood cell numbers to levels that lead to high blood pressure, increased capillary leakage, and edema. The gift of the gene was the result of interbreeding between high altitude

Denisovans and the ancestors of the people who live today on the Tibetan Plateau.[13]

Ancient DNA and protein analysis of Ötzi's stomach contents revealed that he had prepared for his alpine sojourn with a meal of red deer, ibex, and wheat, most likely bread.[14] Preservation of ancient stomach content is a boon to anthropologists, as it allows for studies of the diets of individuals and their paleo-culture. Ötzi's genome showed that he was lactose intolerant and, indeed, his stomach contained no traces of milk proteins. Recent DNA-based evidence suggests that he was not alone in this regard: post-weaning consumption of milk by modern humans in Europe didn't become widespread until about 5,000 years ago.[15] As our ancestors adopted an agricultural lifestyle, other aspects of their diet also changed. An increase in starch consumption was accompanied by multiplication of the number of genes that encode enzymes that degrade starch. Such metabolic adaptations can be monitored through time by ancient DNA research.

Paleodiet research goes hand in hand with the study of the origin of agriculture. Today, the vast majority of current ancient plant DNA research focuses on the origin, evolution, and domestication of food plants that were, and continue to be, essential to human survival. While specimens are invariably from the more recent past, their connection to our everyday lives makes this research of particular interest. In a number of cases, entire ancient plant genomes have been recovered. Ancient crop plant DNA has been detected in 6,000-year-old barley found in a cave near the Dead Sea, 8,400-year-old wheat from a Neolithic site in Turkey, and a 9,000-year-old cereal from an archaeological dig in the Sahara Desert of Libya.[16] Most unusual was the isolation of DNA from scrapings of the inner walls of amphorae (tall jars) found in a 2,400-year-old shipwreck off the coast of Greece.[17] One amphora provided DNA from both olives and oregano. Did the ancient Greeks enjoy oregano-flavored olive oil, or was it merely contamination resulting from the use of the same amphora to store different products?

The birth of research on the origins of New World agriculture is credited to Richard MacNeish, an American archaeologist who worked at the National Museum of Canada. In 1962, he collected kernels of maize from a 5,300-year-old archaeological site in the Tehuacán Valley of Mexico. They sat for decades on a museum's shelf, but eventually, Jazmín Ramos-Madrigal and Nathan Wales sequenced the genome of those kernels, along with those of many other varieties of the plant. As a result of their work, we now know that maize evolved from one of a group of grasses called teosintes in central Mexico. The original plant, unusual for the breadth of its leaves, had a seedpod less than an inch long that contained perhaps a half-dozen small seeds. It still grows in Mexico today.

They also checked the ancient genome against a list of genes known to be responsible for specific morphological characteristics of ancient and modern varieties of maize.[18] One of the genes on the list controls development of exposed kernels—an obviously advantageous property for a crop food; by comparison, each individual kernel of the original teosinte grass was encased in a hard undigestible coat. The Tehuacán Valley seeds were shown to contain the modern version of the gene, an indication that the ancient specimen had already undergone significant selection and evolution by the time it was collected by MacNeish.

Logan Kistler, the curator of Archaeobotany and Archaeogenomics at the National Museum of Natural History in Washington, DC, has traced the evolution of maize through both time and space, just as Svante Pääbo documented the movements and hybridization of the descendants of the Neanderovans.[19] Kistler and his colleagues have sequenced entire genomes of over a hundred different forms of teosinte and its progeny, collected from throughout the Americas. First grown in central Mexico about 9,000 years ago, maize took 1,500 years to appear in Central America and an additional 1,000 years to move to South America, where it quickly spread across the continent as a domesticated crop and became an essential staple for the region, as it remains today.

Ancient biomolecule analysis has taught us about Ötzi's physical health and his diet, but perhaps even more surprisingly, it has also informed us about his fashion. Imagine Ötzi on a catwalk at a fashion show, the commentator's narration describing his accoutrement: "Thigh high leggings, a composite of sheep and red deer skins, loin cloth of goat leather, the coat with broad arresting strips of light and dark-colored sheep and chamois skin, with a fur cap . . ." The clothes that Ötzi wore on that fateful day contained skins and leathers from seven different animal species.[20] In this case, however, the species identifications were not the result of DNA analysis but of keratin fingerprints of the skin and hair.

A Pharaoh's Immunity

Compared to Ötzi, King Tut was a mess. The 19-year-old had a clubbed foot that necessitated the lifetime use of a cane, a nasty fracture of his right knee, scoliosis, and a cleft palate. He had his parents, who were brother and sister, to thank for most of his genetic problems. Nevertheless, he himself married his half-sister Ankhesenamun and, as often occurs in cases of incest, fathered two stillborn children. He also had sickle cell disease and malaria; we know this because of our ability to detect and characterize DNA sourced from both the young pharaoh and his parasites. In a study led by Zahi Hawass, the charismatic former Minister of Egyptian Antiquities, DNA extracted from the mummy of Tutankhamun was found to contain DNA of the protozoan, *Plasmodium falciparum,* the causative agent of malaria.[21] Indeed, researchers detected DNA from two different variants of the parasite, an indication that King Tutankhamun had suffered through two different and distinct bouts of the disease. Despite his high status, he was as familiar with blood-engorged mosquitoes as the stoneworkers who carved his tomb.

The detection of ancient diseases, once a favorite pastime of pathologists with time on their hands, is now an active field of research. Their work has been made much easier by the availability of ancient protein and DNA. Scientists have been able to diagnose deep time cases of leprosy, syphilis, Lyme disease, influenza, the plague, multiple myeloma, sickle cell disease, and a number of different parasitic diseases.[22] In a particularly gruesome example, the lips of exceptionally preserved mummies of a 7-year-old boy and a 15-year-old girl—sacrificed as part of a religious ritual 500 years ago and subsequently buried under half a meter of volcanic ash at the summit of a 22,000-foot-high volcano in Salta, Argentina—were covered with blood.[23] Proteome analysis of a swab of the blood-covered lips of the young Andean girl identified 67 different proteins, several of which were related to inflammation. DNA isolated from the blood sample contained sequences from a bacterium, *Mycobacterium* sp. The girl had suffered severe inflammation due to a pulmonary bacterial infection, most probably tuberculosis.

Inflammation is both a gift and a curse. In response to invasion by a pathogen, cells of our immune system immediately migrate to the point of infection and release what immunologists call a "storm" or flood of defensive signals and proteins. The subsequent cascade of reactions includes increased blood flow and heightened vessel permeability as liquids flow into our tissues. This initial response is meant to be temporary, but sometimes it can do more harm than good. Blood pressure falls as vessels expand and the accumulation of fluids causes swelling. SARS CoV-2, for instance, is sometimes fatal due to the scorched-earth tactics of our immune system. The infected regions of the lungs are identified and encircled. In so doing, these areas are pumped full of fluids and closed off, precipitously limiting the ability of the lungs to provide the body with oxygen.

A surprisingly large fraction of our entire proteome is, in one way or another, involved in the immune system's response to viral infections. The genes responsible for these several thousand proteins are unique in that throughout our evolutionary history they have undergone significant mutation, a testament to the body's continuous attempt to keep up with Earth's huge array of viruses. Neandertals and their ancestors were exposed to influenza, herpes, and HIV-like viruses as they lived in and migrated out of Africa. Recently, extensive genomic analyses have determined that the Neandertals also provided us with a powerful ability to fight these viruses. Prominent among the 2 percent of our DNA that is "Neandertal DNA" are two genes that encode immune cell proteins that constitute a first line of defense—they find and bind to pathogens and signal a call to arms that mobilizes the immune system.

Given the interbreeding between modern humans and Neandertals, many of our ancestors contained, at one point, 50 percent "Neandertal DNA." Over time, Neandertal genes that did not confer an evolutionary advantage were gradually lost. The essential nature of the Neandertal anti-virus genes is indicated by the fact that they have survived in our genome for tens of thousands of years. Our two Neandertal-derived immune system genes encode proteins called toll-like*,[24] receptors. Humans have genes for 10 different toll-like receptors, each of which recognizes and binds to different types of pathogens. Unfortunately, the two genes inherited from the Neandertal genome, while critical to our health, also come with a downside: they are, in large part, responsible for the increased sensitivity some of us show to nonpathogenic materials like pollen. We inherited our allergies from the Neandertals.

Toll-like receptors were also behind Tutankhamun's repeated malarial infections. As with other infectious diseases, the immune system recognizes the malarial pathogen and greets it with an initial flood of defensive proteins that, among other things, induces an inflammatory response. But the malarial parasite introduces a nasty twist to the story. Axel Kallies, Diana Silvia Hansen, and their colleagues in Melbourne, Australia, have determined that interferon, one of the defensive proteins produced in response to malarial infection, interferes with one of the fundamental functions of immune system helper cells (T-helper cells).[25] These cells normally stimulate antibody-producing cells (B-cells) to make antibodies that bind to and initiate destruction of the pathogen. In the presence of the malarial parasite, interferon prevents interaction between these cells, thus preventing the production of antibodies. The disease is unusual in that it can take decades and multiple infections

---

* A perk for scientists is our ability to name new discoveries. Genus species names like *Pieza pi* and *Heerz lukenatcha*, the fruit fly mutants "cheap date" and "Cleopatra" (interaction of the protein products of this gene and the Asp gene result in death), and the subatomic particles quark, strange, and charm are a few examples. The toll-like receptor gene was named after Nobel laureate Christiane Nüsslein-Volhard, who exclaimed, "Das ist ja toll!" meaning "That is weird/amazing!" when she first saw a mutant fruit fly that lacked the toll gene.

before sufficient antibodies are produced to make a patient immune; Tutankhamun never acquired immunity against the malarial parasite.

The practical significance of our Neandertal DNA has been made even more apparent by recent data from the laboratory of Svante Pääbo.[26] There exists a small fragment of Neandertal DNA on chromosome 3 that contains six different genes, one of which encodes a protein on the surface of certain types of immune cells. The production of that protein is correlated with the severity of COVID-19, hospitalization, and respiratory failure. If you have that Neandertal gene and you become infected with the SARS-CoV-2 coronavirus, your chances of becoming seriously ill significantly increase. The Neandertal protein binds to an inflammatory signaling protein that induces the migration of immune cells that participate in the regulation of immune responses in the lungs. The potential mechanisms by which this interaction makes COVID-19 a deadly disease are still being examined.

Neandertal Melanin

There are about 7,000 genetic diseases in humans, maladies in which one or more dysfunctional (or nonfunctional) genes are inherited. Although Linus Pauling was the first to show that a mutation in sickle cell hemoglobin was the cause of a genetic disease, the mutant protein that causes cystic fibrosis was, in 1989, the first disease-causing gene to be cloned. The ability to associate a specific gene or gene sequence with a specific disease spawned the science of molecular medicine. It has also led to a search for genes that, while not directly causative, are in some reliable way associated with not only diseases, but conditions, and even behaviors. For example, there are gene variants that have been shown to be associated with language impairment, a disorder that is to speech what dyslexia is to reading.[27] A paleoanthropologist's thinking might be as follows: If gene X is associated with language impairment in people today, and those exact same genes are in the Neandertal and Denisovan genomes, can we assume that the Neandertals and Denisovans were simply vocalizing, not speaking to one another a hundred thousand years ago? Did the Neandertals have a spoken language? This inference may be a bridge too far, but the ability to find specific genes that, even if their function is unknown, are predictive of a defined trait, like eye color, male baldness, and female breast size, enables us to learn more about our deep time ancestors than was thought possible as recently as a few short years ago.

Pigmentation, which involves dozens of different genes, is one such condition. Pigmentation is a lifesaver in that it protects skin cell DNA from ultraviolet light-induced damage and skin cancer. At the same time, however, that same ultraviolet light is needed to synthesize vitamin D3 from a biomolecule closely related to cholesterol. Synthesis of the vitamin, which mediates calcium absorption in our intestine, may be decreased by as much as 90 percent in the presence of high levels of skin pigmentation. As early species of *Homo* appeared, and as their descendants moved into new and strange environments, the body made compromises, and pigmentation became a highly variable trait. The Inuit, faced with prolonged periods of darkness and cold that required head-to-toe clothing, ended up with low levels of skin melanin. On the opposite end of the spectrum, in the tropical northern coast of Australia, descendants of the last out-of-Africa migration settled and remained isolated for tens of thousands of years in a very warm environment that provided far too much sun.

Gene variants responsible for dark skin color have been shown to have first evolved over 2 million years ago, perhaps when the ancestors of the genus *Homo* lost most of their body hair to produce "the naked ape," as described by Desmond Morris in his book of that name.[28] While most of the known gene variants that favored pigmentation were in place prior to the appearance of anatomically modern humans over 300,000 years ago, some occurred as recently as 29,000 years ago. Gene variants that favored decreased pigmentation, by contrast, originated about 60,000 years ago—at about the same time as the migration out of Africa—and quickly established themselves in Eurasians as a result of positive selection. In one case, analysis of the DNA of a Neandertal from Europe indicated that the individual was not only very light skinned but had red hair.[29]

The genetic basis for this decrease in pigmentation 60,000 years ago may have recently been deciphered by Sarah Tishkoff and colleagues at the University of Pennsylvania. When ultraviolet radiation strikes one of our skin cells, its energy is usually dissipated as harmless heat. If it isn't, it can damage DNA molecules. A nasty sunburn, for example, damages skin cell DNA; if this damage is not repaired, the damaged gene sequence may produce an altered and dysfunctional protein. Tishkoff and her colleagues have studied a number of the genes involved in DNA repair, including a protein named DDB1 (Damage-specific DNA Binding protein 1) that rushes to the nucleus of sunburned skin cells, where it binds to the damaged DNA and initiates repair of the damaged gene.[30] In a beautiful example of a feedback loop, increased amounts of DNA damage signal the skin, via DDB1, not just to repair the damage, but to make more ultraviolet-absorbing melanin. Using molecular clock analyses, Tishkoff's laboratory determined that mutations that decrease production of the DDB1 appeared about 60,000 years ago.

Sourcing Neandertal DNA

Modern research into ancient DNA involves new DNA isolation and amplification methodologies, as well as new sequencing technology. Two decades ago, scientists required years to sequence an entire genome; today, thanks to these new technologies, we can perform the same work in mere hours. But it was a very low-tech discovery—made in 2014 by a team of researchers led by Ron Pinhasi of University College in Dublin, Ireland—that has contributed greatly to the success of modern ancient DNA research.[31] Nearly all ancient DNA is recovered from bone, in part because negatively charged parts of DNA can bind tightly to positively charged calcium in bone tissue. But some bones may be better than others in this regard. Pinhasi's group discovered that the petrous bone, a part of the skull's temporal bone that lies behind the ear, is a superior source of DNA. The petrous bone is very dense, harder than any bone in the body and functions to encase and protect the inner ear. It is such an abundant source of ancient DNA that scientists sometimes partition the history of ancient DNA research into before and after 2014 BP (Before Petrous). But scientists are always looking for something better, and it turns out that the tiny bones of the middle ear itself may be as good or better than petrous bone.

Nearly all of the bones in our body are constantly being remodeled. We have osteoclasts, cells that degrade old or damaged bone, and osteoblasts, cells that make new bone. They are the yin and yang of bone biology, almost always in equilibrium. Without them, our bodies would not have a skeleton, at least not one made of bone. The bones of the middle ear—the stapes, incus, and malleus (commonly called the stirrup, anvil, and hammer)—are the smallest in our body. They are also unique in that they contain few blood vessels and, after one year of age, no longer undergo remodeling—a necessity so that sound transmission, and our ability to hear, doesn't vary. Given the lack of remodeling, existing osteoblasts in the bones of the middle ear become mineralized which, in theory, preserves their DNA.[32] There is a downside though. Isolation of this DNA requires destruction of one or more of the middle ear bones. Luckily, valuable morphological data can be saved by first imaging the tiny bones to create exact three-dimensional replicas.

While the bones of the inner ear may be a gold mine of DNA, like most gold mines, they are difficult to find. Far more commonly found are isolated human teeth. Frequently, as in the Kishenehn Formation in Montana, teeth are the only parts of deep time mammals that are preserved. But teeth too turn out to be a great source of DNA.[33] The enamel that forms the crown of a tooth, produced by cells called ameloblasts, is the hardest structure in the body. As we have previously seen in the case of *Gigantopithecus blacki*, enamel too can be an excellent source of DNA. And below the gum line, the tooth is covered by a layer of densely mineralized bone called cementum, which is produced by cells called, appropriately, cementoblasts. With a collagen content of about a fourth of the weight of the bone, cementum is a good source of both DNA and collagen proteins. Researchers have even found DNA in fossil teeth from the tropics, a place where specimens, because of exposure to very long periods of warm temperatures, have a very poor record of ancient DNA.[34]

It is particularly frustrating when relatively young specimens fail to harbor salvageable DNA. Egyptian mummies have been notoriously difficult in this respect, a result of the chemicals used in the mummification process. Another such case is the hominin skeletons of the island of Flores in Indonesia. First discovered in 2003, the bones of *Homo floresiensis* included parts of 12 different individuals, including a nearly complete skeleton, intact skulls, and a number of teeth. Although the original reports of *Homo floresiensis* suggested that they may have existed as little as 18,000 years ago, more recent data indicates that the youngest skeleton is approximately 50,000 years old; DNA from these specimens should have been low-hanging fruit.[35] However, multiple attempts to isolate DNA from these bones have failed. What makes the absence of DNA so dismaying is the huge scientific potential that resides in the *Homo floresiensis* genome.

Homo floresiensis was unusual. Short, perhaps three and a half feet tall, with a brain a little more than a third the size of ours and a notably sloped forehead, *Homo floresiensis* inhabited the island for hundreds of thousands of years with no contact with the modern *Homo sapiens*, who were busy island-hopping to the north and west as they made their way to New Guinea and on to Australia. Scientists believe that when *Homo floresiensis* first arrived on the island, they were much larger, but over time grew smaller due to a lack of resources, a phenomenon termed island dwarfism. In fact, one of *Homo floresiensis*'s primary food sources was a pigmy elephant, its small stature the result of the same phenomenon.

By sequencing the Neandertal and Denisovan genomes, in addition to that of modern humans, scientists have been able to identify particular sequences as having originated with and once been unique to each of these three species. They have also found "ghost" sequences in these three genomes, which are thought to be derived from a "superarchaic" common ancestor of the Neanderovans and *Homo erectus* or even *Homo erectus* itself.[36] Recently, scientists have suggested that *Homo floresiensis* may have diverged directly from *Homo erectus* and a lineage largely unrelated to that of the Neanderovans.[37] If we possessed the genome of a *Homo floresiensis*, it could tell us about the origin of this enigmatic hominin and perhaps even the ghost sequences. And not just their origins. With an entire genome in hand, we might even be able to determine the genetic basis of their short stature.

In a recent study, the genomes of 10 individuals of a population of pygmies that currently live on Flores were sequenced. Both Neandertal and Denisovan gene sequences were found but nothing to indicate a potential connection with more ancient lineages, such as the extinct *Homo floresiensis* and their ancestors. However, the pygmy genomes were significantly enriched in multiple gene variants that are, in *Homo sapiens*, associated with reduced height. It appears that the pigmies that live on Flores today are descendants of modern humans who, isolated on a small island, were subject to the same environment-based selection processes that befell *Homo floresiensis*. Hopefully, one day soon, a team of scientists will harvest *Homo floresiensis* DNA. But if they can't, they will have a well-documented excuse: a warm climate.

Ancient DNA does not deal well with heat. There are exceptions of course (for example, the 360,000-year-old cave bear discussed in the previous chapter), but almost all ancient genomes have been recovered from fossils found in the northern hemisphere. The average minimum temperature on the island of Flores is 72 degrees Fahrenheit (22 degrees centigrade). It has been that way for the past 15,000 years, and was only very slightly cooler during the proceeding 55,000 years. Although the first skeleton of *Homo floresiensis* to be found in Flores was buried under 20 feet of soil, the temperature gradient relative to soil depth in the tropics is very shallow; it simply is not cold enough to preserve DNA. Perhaps, however, ancient proteomics will provide data where DNA analysis can't. Using both technologies, and with new and more sensitive technologies in the offing, anthropological and archaeological research are entering an era of historic expansion, based on ancient biomolecules.

11

PLANTS

Research on ancient genomes and proteomes is radically expanding and forcing us to revise our understanding of the past. Unfortunately, this may be of little use in our attempts to understand the phylogenies and ancient physiologies of the vast majority of life on Earth, at least as measured by total biomass. The biomass of Earth's plants is more than 15 times as great as that of all the planet's multicellular animal life combined. As documented in chapter 9, the fossil record of in situ DNA and proteins in deep time plants is nonexistent. Yet the lack of DNA and proteins does not exhaust the universe of ancient biomolecules. Consider, for instance, the 558-million-year-old cholesterol-like molecule isolated from a specimen of the Ediacaran *Dickinsonia*, which we encountered in chapter 2. That ancient biomolecule is small, essential to life, and its structure completely understood. Any biochemist could quickly draw its structure on a whiteboard in a few seconds. And how about huge cross-linked and derivatized polymeric molecules, many of which modern science doesn't fully understand? Sporopollenin, for example, a major component of pollen, fits this description well. Taking into account such biomolecules, the majority of ancient biomolecules are, by far, from plants.

Some of the oldest known organisms with a nucleus, the eukaryotes, include the 15 different 1.6-billion-year-old species that were collected north of Beijing, China, by Maoyan Zhu at the Nanjing Institute of Geology and Palaeontology.[1] The fossils themselves are neither impressions in rock nor the compressed remains of intact ancient organisms. They are, instead, the organisms' external walls, which are composed entirely of ancient organic biomolecules (which is to say, they have not been permineralized). Much of our understanding of deep time life is based on the study these organic-walled microfossils, often referred to as OWMs. Some of the OWMs described by Zhu's team of researchers, including *Valeria lophostriata*, *Schizofusa sinica*, and *Cucumiforma* sp., look much like microscopic versions of the dried paper-like outer covering of a tomatillo. Their walls are composed of what is often understatedly referred to as a "recalcitrant organic material." Although the term doesn't tell us much about the material's composition, it encompasses much of what, until very recently, we knew about it. It's also a bit of an oxymoron. Original organic biomolecules, which we usually think of as exceedingly labile and the first thing to disappear during the fossilization process, are responsible for some of the oldest fossils known to science.

OWMs also include fossils of microbes, including a 3.2-billion-year-old fossil found in South Africa, one of the oldest fossils on Earth.[2] Such OWMs are assumed to be cysts of bacteria and spores of fungi, unique structures that allow organisms to remain dormant until conditions are favorable for germination and renewed growth. One can imagine how tough the exterior of a spore must be for it to simply rest in place, buried in ocean sediments for years, and avoid destruction. Although permineralized fossil pollen has been recovered from sediments over 300 million years old, organic-walled pollen is not far behind, with collected specimens at least as old as the early Jurassic.[3] The organic walls of these various structures, both ancient and living, are composed of a variety of very complex materials, many of which have yet to be fully characterized.

As discussed in chapter 2, a crucial piece of evidence in studies of the Chicxulub asteroid apocalypse that killed off the dinosaurs is the boundary layer that contains remnants of the asteroid's impact, marking the end of the Cretaceous period. Antoine Bercovici, an exceedingly amiable Frenchman and palynologist (someone who studies pollen), has spent the last decade studying the ancient sediments below and above the Chicxulub boundary layer. Bercovici was able to identify numerous pollen species, referred to as K (Cretaceous)-taxa, whose existence terminated at the layer. In the sediments produced after the asteroid impact, large numbers of a single species of fern spores were found.[4] Their presence makes perfect sense—ferns are recognized as early colonizers or disaster flora, plants that are the first to appear after an event such as floods, lava flows, or ash deposits that denude the landscape of all life.

The preservation of both the K-taxa and the disaster flora is due to sporopollenins, huge biomolecules—let's call them macrobiomolecules—that form a tough, resistant exterior sheath around both pollen grains and spores. The term "sporopollenin" was coined in 1931, but its initial identification was based more on its functional characteristics than its chemistry. It is resistant to treatment with hot solvents, corrosive alkali, and acids. Vera Korasidis, a palynologist postdoctoral fellow at the National Museum of Natural History, allowed me to watch her isolate fossil pollen from 65-million-year-old mudstone—a procedure that offers testimony to the tough nature of the sporopollenins and explains why most of her colleagues prefer to avoid such work.

When we entered the pollen isolation laboratory, the first item on the agenda was personal protective equipment: rubber boots, plastic arm-length gloves, safety glasses, and, over those, a face shield; we then donned a lab coat and heavy apron. Korasidis turned on the ventilation system and picked up a test tube of rock, which she had earlier crushed to a sand-like consistency. The protocol varies depending on the sample. Coal, a great place to find pollen, can be processed in just a single day. But the mudstone from Wyoming that we were processing would require nearly four days of treatments with numerous nasty reagents. The crushed rock was first treated with hydrochloric acid, which dissolves carbonate rocks like limestone. Next up was a 24-hour treatment with the exceedingly dangerous hydrofluoric acid reagent, which dissolves silicate-based rocks like quartz. Korasidis related a story about a colleague who, in the process of isolating fossil pollen, was using hydrofluoric acid in a ventilated hood. Unbeknownst to him, the hood was not working. He left the lab with multiple holes that penetrated completely through his contact lenses. While his eyes were thankfully not permanently damaged, we decided to double-check the hood, just to be safe. The specimen was then subjected to subsequent rinses with hydrochloric acid and water, sieving, another treatment with hydrofluoric acid, and incubations with nitric, acetic, and sulphuric acids, the latter two designed to remove residual plant debris, such as cellulose. It was a laborious process, but Korasidis was rewarded with an ancient botanical bonanza: her sample provided more than a hundred different species of pollen.

Sporopollenin, as the title of one scientific paper has aptly stated, is "the least known yet toughest natural polymer."[5] It's a complex, heterogenous, cross-linked polymer that is highly resistant to bacterial degradation, as well as to temperatures as

high as 300 degrees centigrade (572 Fahrenheit). Even insects
that depend on pollen for food can't degrade it. Honeybees
swallow pollen grains whole, and once in the insect's gut, the
pollen grain is shattered by osmotic shock, the nutritious core
of the pollen is exposed and digested, and the tough outer
sporopollenin coat is left behind. Only very recently has the
complete structure of one of the sporopollenins been elucidated. The work of Jing-Ke Weng and his colleagues at the Massachusetts Institute of Technology's Whitehead Institute have
shown that the structure of pine pollen is even stranger than
previously imagined.[6] If you really need to know—and you
don't—it's made of "aliphatic-polyketide-derived polyvinyl alcohol units and 7-O-p-coumaroylated C16 aliphatic units cross-
linked through a distinctive m-dioxane moiety." There will be a
quiz at the end of the chapter.

Given its near indestructibility, the fossil record of the
molecule itself is excellent. Barry Lomax at the University of
Nottingham in the United Kingdom and his colleagues have
identified and analyzed sporopollenin in 310-million-year-old
spores from *Rotatisporites dentatus*, an extinct species of lycophytes (for example, clubmoss).[7] Structural characterization of
the ancient biomolecule demonstrated its near identity to sporopollenin found in spores of lycophytes that live today. The
indestructible nature of sporopollenins and the ease with which
pollen from living trees can be isolated in large quantities provides
another instance in which biomolecule research has commercial potential. A company called Sporomex purifies the organic
walls of pollen, which it uses to manufacture a product that
serves as a vehicle for the delivery of pharmaceuticals. The final
product is resistant to stomach acids, is nonallergenic, and, as the
active agent is slowly released through pores in the pollen's outer
coating, can provide a timed-release dosage of the drug.

The list of the complex polymeric macrobiomolecules of plants that are found in the fossil record is a short one, but these biomolecules are critical to the existence of most organisms on earth. Without them, dormant spores would never germinate, trees would not grow tall, and leaves would dissolve in the rain. To understand how these biomolecules can shed light on ancient (and perhaps future) life on this planet, let's examine three of them: cutin, cellulose, and lignin.

Cutin and Climate Change

As a Peace Corps volunteer in Samoa in the early 1970s, I loved to hike into the untouched tropical forests, which contained the most diverse flora I had ever seen. One weekend I hiked inland from the school (at which I taught science and math) to Mount Matavanu, about seven miles inland from the ocean. On the way back from the long and hot walk over black lava fields, I was met by a young man from the village of Fagamalo. It had just started to rain, and I tilted my head back in a fruitless attempt to catch a drink of water. The young Samoan saw this and lifted the leaf of a banana tree at the side of the trail and bent it into the form of a large V-shaped vessel. Within no time, I had a liter of water to drink, thanks to the kindness of a stranger, and more importantly for our purposes, another complex plant-specific macrobiomolecule: cutin.

This biomolecule, which makes the surface of leaves hydrophobic, is a major component of the exterior cuticle of leaves and functions to form a protective barrier and prevent water loss. A waxy material, it is a polymer that is huge and ridiculously complex in its composition. It is also impressively represented in the fossil record. Ancient cuticles, little changed by fossilization and containing ancient cutin biomolecules, have been found in fossil plants approximately 400 million years old. Amazingly, fossil leaf cuticles themselves have also been recovered, intact, from fossils hundreds of millions of years old. Cutin itself has been purified from cuticles of a 305-million-year-old fossil seed fern.[8] Like cutin purified from the leaves of a living tree, the purified material was yellow.

In potentially one of the most important studies involving ancient plant biomolecules, Richard Barclay is attempting to use both ancient and present-day cuticles to shed light on the future of our planet by turning to the deep past. Barclay, a geologist and paleobotanist at the Paleobiology Department of the National Museum of Natural History, is trying to determine the temperatures that prevailed on Earth hundreds of millions of years ago. On an acre of land adjacent to the Smithsonian's Environmental Research Center (SERC, as it is commonly known), containing little other than a couple of abandoned greenhouses, Barclay has used two-inch-diameter PVC pipe and sheets of transparent plastic to erect numerous structures that are eight feet on a side, twelve feet high, and open at the top. Within each is a single *Ginkgo* tree.

The leaves of these trees, like all plants, take up carbon dioxide (CO_2) and give off oxygen through openings on their surfaces, called stomata, that they can open and close. In the process, however, they unwittingly allow water to escape. This is a problem, as regulation of its water content is critical to the health of the plant. Barclay's research is based on two observations in the scientific literature. The first is that the number of stomata is inversely proportional to the CO_2 content of the atmosphere. If the CO_2 content is high, the plant requires fewer stomata to take in the CO_2 needed to produce sugars via photosynthesis. As the concentration of CO_2 increases, plants actively decrease the numbers of stomata in emerging leaves. This is advantageous to the plant because with fewer stomata, less water is lost. The second, and much more widely known observation, is that our planet's surface temperatures are related to the carbon dioxide (CO_2) content of the atmosphere. The higher the CO_2, the higher the temperatures.

Barclay's research at SERC is designed to test the first of these two observations in a long-term real-world setting. Do leaves produce fewer stomata as temperatures increase? If so, perhaps the numbers of stomata in fossil leaves could be used as a proxy for deep time temperature.

When I drove to SERC to speak with Barclay, the first person I met was Mike, the driver of a tanker truck full of CO_2. Mike had just delivered 2,400 pounds of the gas, a run he makes on a weekly basis. The CO_2 is pumped into Barclay's plastic structures to provide air with CO_2 contents of 400, 600, 800, and 1,000 parts per million. A group of volunteers helps to maintain the 12 structures, monitor their CO_2 content, and make measurements of temperature, relative humidity, rainfall, and so on.[9] Initiated in March 2016, and with years to go until completion, this is not the type of experiment to do if you want to pad your publication record.

Volunteers also count the numbers of stomata in leaves taken from the *Ginkgo* trees. The leaves are covered by a cuticle, which can be isolated intact, with well-defined outlines of the stomatal cells. For just a moment, let's put ourselves in the place of Sal Bosco, an affable retiree and "gym rat" in his seventies who volunteers with the project. Sal's job is to treat one-centimeter-square pieces of *Ginkgo* leaves with chemicals that remove almost all of the carbohydrate-based organic material (mostly cellulose) between the upper and lower cuticles. After treatment, two transparent cuticles remain, one from the top of the leaf and one from the bottom, like two small pieces of transparent

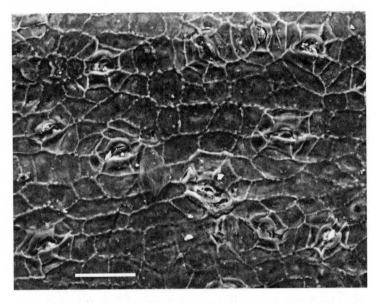

FIGURE 11.1. The stomata of a fossil *Ginkgo* leaf stand out when viewed through a scanning electron microscope. Scale bar = 50 microns.

tape adhered one to the other. Observing the material through a microscope, Sal separates the two layers of cuticle with the aid of fine forceps and identifies the upper and lower layers and the inner and outer sides of each. The isolated cuticles tend to curl up like a diploma scroll, so it's a job that few volunteers have the patience to undertake. Once separated, the lower cuticle must be spread on the surface of a sample holder for subsequent analysis.

So why the *Ginkgo* tree? Barclay chose this particular tree because it is a living fossil. The genus has been around for at least 201 million years, it has an excellent fossil record, and, astonishingly, the cuticle of fossil *Ginkgo* leaves, with their constituent cutin molecules, can be isolated intact and their stomata counted.[10] Fifty-five million years ago, global temperatures are thought to have spiked to 8 degrees Celsius (14 degrees Fahrenheit) above those that we observe today.[11] Can the CO_2 content of the atmosphere that existed 56 million years ago be determined through the use of fossil *Ginkgo* leaves? If so, can our estimates of the temperatures that existed in the Cretaceous be confirmed? Such information would be an invaluable addition to our knowledge of past environments, as well as enhance our ability to predict future temperatures as we continue to add CO_2 to today's atmosphere.

250-Million-Year-Old Cellulose?

Cellulose is the most abundant biomolecule on the earth's surface. It has been around for a very long time, probably much longer than one very conservative estimate that places it at the beginning of the Cambrian, 541 million years ago.[12] This estimate is not based on the preservation of ancient cellulose per se, but rather on a huge increase in carbon deposits that occurred at that time. An even older age for the appearance of cellulose is supported by several additional observations. Cellulose is a structural element of red and green algae, both derived from a common ancestor that lived more than 600 million years ago. Enzymes that degrade cellulose are present in nearly all fungi, a taxon that has been on earth for approximately 1.5 billion years, and enzymes that make cellulose are present in cyanobacteria, an even older group, fossils of which are 3.5 billion years old.

A polymer composed of 10 thousand or more molecules of a derivatized form of glucose, cellulose makes up more than 40 percent of the content of dried wood. Like many of the ancient macrobiomolecules, it is long-lasting. In a study of ancient plant material collected from Axel Heiberg Island in the

Canadian Arctic Archipelago, scientists calculated that over a quarter of the plant tissues' original cellulose was still present.[13] In a more recent study by Leszek Marynowski at the University of Silesia, small fragments of 13-million-year-old wood were found to be 55 percent cellulose.[14] One of the few ancient biomolecules that have been purified in relatively large amounts from fossil plants, ancient cellulose has been purified from both 11-million-year-old wood from a lignite mine in Germany and 47-million-year-old *Metasequoia* stumps from Axel Heiberg Island.[15] The purified product is a slightly yellowish off-white fibrous material that looks something like a cotton ball.

While this ancient wood is indeed a scientific gold mine to a paleobotanist, to other scientists, it is merely wood: a colleague at the National Museum of Natural History tells the story of an expedition of vertebrate paleobiologists to Ellesmere Island in the Canadian Arctic Archipelago, where it can be uncomfortably cold even in the summer. To feed their campfire, the scientists gathered up and burned pieces of a 47-million-year-old dawn redwood (*Metasequoia*) tree. A paleobotanist listening to the recitation, thinking perhaps of the single specimen of *Metasequoia* in the museum's collections, was justifiably aghast.

Hints of ancient cellulose have even been found in wood from Poland that is 168 million years old. Surprisingly, the diagnostic degradation product of cellulose was found in the rock-hard permineralized inner portion of the fossil tree.[16] Perhaps we should look at the trees of the Petrified National Forest a bit differently the next time we visit the park. Without question, though, the most spectacular report of ancient cellulose comes from the Salado Formation, located in the southeastern corner of New Mexico.[17]

Deposited 253 million years ago, before the appearance of dinosaurs, the Salado Formation contains numerous thick layers

FIGURE 11.2. Branches of 47-million-year-old *Metasequoia* wood from the Canadian Arctic Archipelago retain tunnels made by wood-boring beetles. Scale bar = 4 inches.

of salt (sodium chloride, or rock salt), a discovery that must have been a surprise to the wildcatters who drilled for oil and gas in the area many decades ago. The layers of salt crystals originated as concentrated oceanic brine on the surface that, over time, seeped into the soil and was eventually buried by millions of years of deposition. Recovered at a depth of nearly a half-mile, the largest of the crystals were about four inches in length. Many of them contained tiny liquid inclusions, brine that had not become part of the crystal structure itself. Using a microbit, scientists were able to drill through the salt crystal and into the inclusion to harvest the liquid. Suspended in the inclusions of brine were beautiful intact fibers.

Attempting to identify the chemical nature of the fibers, researchers heated them in an alkaline concoction that destroys almost all biomolecules. The fibers were unaffected. The scientists knew that cellulose is one of the few biomolecules that is resistant to this particular alkaline brew, so they performed a more refined chemical test. Although humans can't digest cellulose, many bacteria (like those in a cow's stomach) and fungi are able to do so with an enzyme called cellulase. (The suffix "-ase" is commonly used to name enzymes that cut up biomolecules; sucrase cleaves sucrose, proteinases cleave proteins, and chitinases cleave chitin.) When the fibers were incubated with cellulase, they were digested and nearly disappeared. Based on this evidence, the scientists reported that the salt crystals contained ancient cellulose fibers.

The story garnered much more attention when the same team of scientists subsequently published a paper that described the recovery of a bacterium—belonging to the extant genus *Bacillus*—from the same salt crystal inclusions as those reported to contain cellulose fibers.[18] Not only did they recover a bacterial cell, but it was alive; the investigators were able to grow the bacterium in culture. Talk about de-extinction! The paper was published in *Nature*, arguably the most influential of all scientific journals, but was soon followed by several reports that were highly critical. The laboratories that reported both the cellulose fibers and the living bacterium had taken extreme measures to prevent contamination of their samples; in this case, contamination was not an issue. Instead, the criticism centered on the age of the salt crystals themselves—were they really 250 million years old?[19] One particular molecule of DNA from the recovered bacterium differed from that of a very closely related species of living bacterium by only three mutations. Over a period of 250 million years, given a constant mutation rate, the sequence differences between the supposedly ancient bacterium and extant bacteria should have been 20 times greater. At that point, pretty much everyone concluded that the most parsimonious explanation was that recent water, carrying recent bacteria and cellular remnants such as recent cellulose fibers, seeped into the ground and formed new salt crystals. Cellulose fibers preserved for 250 million years? In this case, no. But the research served to affirm the value of scientific criticism by one's peers.

Ancient Lignin

Although cellulose is the most abundant macrobiomolecule on our planet, lignin is not far behind. In higher plants, lignins (the Latin word *lignum* means "wood") are huge, complex, cross-linked polymeric organic biomolecules found in and around the long vessels of the plant's water-conducting system. Lignins are hydrophobic, so their presence makes the inner surface of the vessels impermeable to water, thereby allowing the long-distance transport of water and nutrients. Lignin also reinforces plant cell walls and, along with cellulose, makes them rigid. These two features—narrow water-impermeable vessels and structural rigidity—are essential to trees: without them, trees would be unable to grow much beyond ground cover. This observation is important because it may provide us with a proxy for the date of the first appearance of lignin itself.

In 2007, William Stein and his colleagues discovered a fossil related to today's horsetails and ferns in 385-million-year-old exposures near Gilboa, New York.[20] It consisted of the crown, base, and trunk of a tree about 26 feet tall. The height of the tree alone provides strong evidence, indirect though it is, that lignin biomolecules were present more than 385 million years ago, and probably much farther back in time. Ancient lignin molecules themselves have been recovered, intact, from deep time fossils that were found in a seemingly unlikely place. Back in deep time, the Canadian Arctic Archipelago was much warmer than it is today, and the islands of the archipelago are known for their deposits of fossil leaves. Often described as made of "leaf litter," some deposits contain wonderfully preserved conifer leaves, with epidermal cells and wax cuticles, that date from the Paleocene to the Miocene. Ancient lignin has been identified in both 45- and 60-million-year-old *Metasequoia* (ancient redwood) leaves from Axel Heiberg and Ellesmere Islands, respectively.[21] Some of the fossil woods were more than 80 percent lignin. This large structural biomolecule has also been found in *Metasequoia, Magnolia,* and *Quercus* (oak) leaves from the 15-million-year-old Clarkia site in Idaho.[22] If you collect at the Clarkia site in Idaho, keep in mind that the ancient leaves that you will hold in your hands contain lignin, as well as cellulose and chlorophyll, and, as we will now see, a host of much smaller biomolecules.

In past chapters, the potential for ancient biomolecules to inform us about the phylogeny, physiology, pigmentation, and behavior of deep time organisms has been a primary and eye-opening theme. With plants, not so much. Cutin, cellulose, and lignin, all large structural molecules, are found in almost all plants and serve nearly the same function wherever they occur. But plants also produce a vast array of smaller molecules that are unique to the plant kingdom, including several with which we are intimately familiar. Some are quite nasty, like the rash-inducing oils of poison ivy and poison oak and the sticky yellow resin that seeps from the damaged bark of a pine tree. Many are bitter and potentially poisonous, such as caffeine, strychnine, quinine, mescaline, nicotine, cocaine, and morphine. On the other hand, what would we do without peppermint, eucalyptus, lavender, clove, oregano, and rosemary oils? The list of these smaller molecules is a very long one, but it is worth discussing a few of the more interesting scientific findings from this expansive universe of extant and ancient biomolecules.

Chemotaxonomy

If you are a morning tea or coffee drinker, you begin each day by making a complex extract of plant leaves or seeds. But do you know their chemical constituents, other than caffeine? Analyses of these extracts detects as many as 82 (tea) and 77 (coffee) different chemicals—and these are just the volatile ones.[23] There are, in fact, over 100,000 such compounds that are made by plants. And many of them can be found in fossils. Their presence can tell us much about the physiology of deep time plants and even the behaviors of the organisms that interacted with them. Why do more than 60 different plants synthesize caffeine? Like many plant-specific small molecules, caffeine acts as a natural pesticide that paralyzes and kills insects that feed on their leaves. Similarly, citrus oil from the peel of an orange is a volatile and potent insecticide. It repels insects and dissuades them from burrowing into the external surface of oranges and related fruit. If you have any oranges on hand, you can do a little experiment that will demonstrate the chemical nature of the oil. Squeeze an orange peel to the point where the external skin breaks, and a small geyser of volatile molecules will erupt. Hold a match to the discharge, it will burn, albeit briefly, like a torch. The question for the paleobiologist is, as always, can we detect these and other compounds in situ? We can, to a point where the detection of these compounds has formed the basis of a new subfield of taxonomy.

Chemotaxonomy is based on the existence of small organic molecules that are taxon specific. Cannabis produces different chemicals than lettuce; *Vitis vinifera,* the common grape vine,

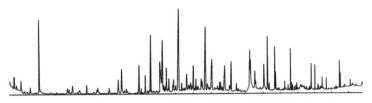

FIGURE 11.3. The array of volatile small molecules in an extract of coffee. Each peak represents one unique component.

synthesizes different chemicals than oak trees. Family-specific, genus-specific, and even species-specific small biomolecules exist. We can, in theory, classify plants based not on examination of their morphology or DNA, but on their repertoire of small biomolecules. Chemotaxonomy has been around for decades, and its application to living plants has demonstrated its scientific legitimacy. For example, in a study of 17 different species of Japanese chrysanthemums, the presence of specific small biomolecules divided the various species into three distinct groups, a division that largely agreed with groupings based on morphological characteristics alone.[24] Some individual species could even be differentiated based on the presence of a single type of small organic chemical. But can the concept of chemotaxonomy be applied to deep time fossils?

The biomolecular components of plants have been found as biomarkers in rocks that are two and a half billion—with a "b"!—years old.[25] They are, of course, no longer physically associated with fossils. These include many of the plant-derived small chemicals that make up amber, its chemical components having been long separated from the resin-producing tree that was its source. The oldest amber, buried in an Illinois coal deposit, dates back to about 320 million years ago, its components cross-linked and polymerized to a near glass-like consistency.[26]

Scientists have coined a term, paleochemotaxonomy, for the classification of fossil plants based on the in situ preservation of unique arrays of small organic molecules. Plant resin-derived biomolecules have been isolated from 168-million-year-old wood found in Poland, 45-million-year-old dawn redwood trees from the Arctic, and 15-million-year-old seed cones of bald cypress trees found at the Clarkia site.[27] (You will recognize these sites from previous chapters; since the study of ancient biomolecules is a young science, researchers are still working at a relatively small number of sites.) Imagine the potential for the identification of plant fossil remnants too poorly preserved to identify based on morphology. All one would need to do is make an extract and examine its array of small organic chemicals.

Earlier in the book, we saw how the original assignment of the Burgess Shale sponge *Vauxia* to the taxonomic subclass Keratosa was confirmed by the detection of the ancient biomolecule chitin in *Vauxia* fossils. The use of ancient biomolecules to make taxonomic assignments is much more common in plants. Consider, for instance, a recent study of ancient leaves from beech trees (genus *Fagus*).[28] Having discovered a fossil leaf and an attached fruit that appeared to be related to living trees of that genus, but which displayed a distinct morphology, researchers placed the fossil in a new genus, *Pseudofagus* (the Greek word *pseudo* means "false"). When scientists later examined their repertoires of small organic molecules, the two species could be distinguished by the presence, in *Pseudofagus* only, of several unique compounds, including a steroid similar in structure to cholesterol; paleochemotaxonomy had confirmed the initial, morphology-based taxonomic assignment.

Despite its potential, chemotaxonomy has not been embraced by the paleobiology community. Their resistance may be related to the dictum that "absence of evidence is not evidence of absence." Many small organic molecules are notoriously unstable at elevated temperatures. Their absence in a fossil does not prove that they were not once present. And degradation of small biomolecules into numerous different, smaller derivatives can be a problem if their presence is overinterpreted. This is exemplified by the work of scientists at the Kunming Institute of Botany in China, who, having isolated numerous small organic molecules from 120-million-year-old ginkgo (genus *Ginkgo*) tree leaves found in a coal mine in Inner Mongolia, wondered about their significance.[29]

In addition to examining the ancient biomolecules of the fossil leaves, they isolated small biomolecules from both a leaf of a living ginkgo tree and leaves of the same species that had been archived in a botanical collection for 150 years. The profiles of small molecules from the three specimens differed greatly. When the data were quantified, they revealed 33 small organic biomolecules in the deep time fossil, 21 in the 150-year-old leaves, and only 13 in the fresh leaf. The older the specimen, the more it degrades, and the more fragments of the original molecule there are with which to contend. The biomolecules unique to the fossil leaf did not necessarily provide evidence for a new species of ginkgo. In some cases, however, the paleochemotaxonomist's job is made easier if enough degradation products have been preserved to allow identification of original parent biomolecule(s). Doing so, however, will require more collaboration between paleobotonists and chemists.

Beyond helping scientists make taxonomic determinations, ancient plant biomolecules can also be used to trace behaviors—not of plants, but of humans. Dr. Patricia Crown of the University of New Mexico has spent much of her career studying the archaeology of the American Southwest. She and her colleagues had long suspected that the distinctive ceramic drinking vessels found at Pueblo Bonito in the Chaco Canyon National Historical Park, like similar jars used by the Mayans, may have been used to consume chocolate.[30] To test whether this was in fact the case, her team ground up small pieces of the inner surfaces of the ceramic jars, made extracts of the powder, and analyzed them with mass spectrometry. In order to prevent the lab's coffee drinkers from contaminating the 900-year-old samples—caffeine is a component of cacao—the scientists took a pointer from modern ancient DNA methodology and wore masks, gloves, and gowns. Cacao contains several bitter alkaloids, primarily theobromine, at a concentration several times that of caffeine. The alkaloid is similar to caffeine but lacks the latter's ability to affect the central nervous system. Brown's group found the several different alkaloids character-istic of cacao and, in doing so, advanced our understanding of the trading that occurred between the peoples of southwestern North America and Central America.

While the study of ancient small biomolecules in plants has not achieved the same level of recognition as the study of ancient DNA, we should not be too quick to write off its potential. As new and more sensitive analytical techniques are developed, perhaps only microscopic fragments of irreplaceable fossils will need to be sacrificed to perform analyses. Perhaps algorithms will be developed that will accurately interpret the unique fin-gerprints of the fossilization-dependent degradation products of ancient plant biomolecules. Given the technical advances of the past few years, the future looks very promising.

12

THE FUTURE OF STUDYING THE PAST

As I hope this book has demonstrated, the current science of ancient biomolecules looks dramatically different compared to even 10 years ago. We have witnessed an explosion of progress, such that few outside the field are aware of how far we have come in such a short time. In this final chapter, we cast our eyes toward the future of ancient biomolecules and contemplate what intriguing discoveries might be on the horizon. Will ancient genomes allow us to produce viable embryos and clone long-extinct animals? Will we be able to make proteins that existed billions of years ago? Can ancient biomolecules help answer the question that we as *Homo* have been asking for millennia: When did life originate on Earth?

When it comes to life on other planets, scientists are already busy looking for ancient biomolecules. In 2012, the 10-foot-long rover *Curiosity* landed just inside Gale Crater on Mars and began its ponderous descent over buttes and through Martian river channels. This area was selected for exploration in part because the channels and deltas suggested the former presence and flow of water. And, sure enough, soon after landing the rover photographed vast areas of beautifully layered mudstone sediment that seemed to confirm that the crater had once contained a huge lake. It wasn't until several years later, however, that *Curiosity* drilled into the lake bed and recovered rock from two inches below the surface. Crushed to a fine dust, the samples were analyzed by a technique called pyrolysis, which involves heating them to as high as 1,500 degrees Fahrenheit. The extreme heat breaks chemical bonds, which results in the release of small volatile molecules for analysis. The scientists at NASA had literally hit pay dirt.

The *Curiosity* rover identified dozens of organic compounds, including propane and benzene with its classic ring of six carbon atoms. The rover also discovered an abundance of organic molecules that contained sulfur, as well as nitrates, a possible source of the nitrogen that is found in biomolecules such as DNA.[1] Both elements are essential to life on Earth— proteins would not be proteins without their nitrogen and sulfur. The problem with pyrolysis, of course, is that it allows us to glimpse only fragments of molecules. What were the much larger parent compounds buried two inches deep in Martian soil?

We will soon have an answer to this question, as *Curiosity* is being joined by a flotilla of newer, more technically advanced rovers capable of detecting Martian biomarkers. For instance, on February 18, 2021, a rover named *Perseverance* landed in Jezero Crater, an area that is 3.5 billion years old and 30 miles in diameter. Multiple lines of evidence suggest that this crater once held a freshwater lake. This evidence includes photographs of the crater that clearly show its water source, a large mudstone-rich river delta, and subsequent photographs of obviously layered rocks. *Perseverance* recently conducted its first analyses of Martian rocks, and NASA scientists were rewarded with even more proof that Jezero Crater once held water. The rover identified these rocks as basalt, volcanic in origin. However, they also discovered, within the basalt, small crystals of salts such as calcium phosphate and calcium sulfate. The latter salt is known on Earth as gypsum, a mineral that forms as a result of evaporation of mineral-laden water. NASA scientists think that Martian groundwater percolated through the porous basalt and, as it evaporated, the minerals formed crystals. Even more tantalizing is the possibility that the crystals trapped tiny bubbles of the ancient Martian water itself.

The two American rovers will soon be joined by *Tianwen-1*, the first rover from China, and in 2023, a rover sent by Russia and the European Union that is named after Rosalind Franklin, one of the scientists responsible for the discovery of DNA.

As with most scientific endeavors, the data produced by *Curiosity* gave rise to more questions than answers. The compounds identified by that rover need not have been derived from living organisms. The sulfur identified by *Curiosity*, for example, likely came from the abundant sulfate minerals found in the soil; might the organic compounds also have originated from geologic phenomena, such as a Martian volcano, or might they have been transported to Mars by a meteorite? Or perhaps they are some form of kerogen, the major plant-derived organic component of oil shale here on Earth. Perhaps. The layers of sediment in the Gale Crater are thought to be about 3 billion years old—plenty of time for evolution of life, especially given the strong evidence for the existence of ancient lakes and streams. However, at present, there is currently no evidence that any living organism has ever been to Mars—a fact that NASA bent over backwards to convey when they released *Curiosity*'s data to the public.

Are authentic ancient biomolecules on Mars so implausible?[*,2] The cyanobacteria that produced 3.7-billion-year-old stromatolites in southwestern Greenland are as old as the rocks at Gale Crater, and the fossils in the 3.43-billion-year-old sandstone of the Stelly Pool Formation from Western Australia are nearly so. It is entirely possible that the *Curiosity* rover is slowly travelling past rock that contains microscopic Martian fossils; *Curiosity* simply lacks the capability to find them. The arrival of the new generation of rovers, as well as future missions to Mars, which hope to transport rocks back to Earth, may allow scientists to overcome this limitation and answer one of the most intriguing questions about our neighboring planet.

It's not just Martian life that is controversial. Scientists are still arguing over the Apex locality fossils. For decades, fossils of thermophilic filamentous bacteria from the 3.46-billion-year-old Apex Basalt in Western Australia were thought by some to be the oldest evidence of life on Earth. However, to call that interpretation contentious would be an understatement. In 2015, David Wacey at the University of Western Australia and his colleagues, using state-of- the-art three-dimensional nanoscale microscopy, proposed that the filaments were nothing more than silicate minerals covered with a thin film of carbonaceous material, all of geological origin.

At the same time, however, this same team of scientists provided evidence that the only very slightly younger rocks of the Stelly Pool location were in fact coccoidal bacteria. Wacey and his colleagues were not alone in their skepticism: a recent report

* Unlike the putative biomolecules found on Mars, which could be preserved remnants of ancient life, the recently reported phosphine biomolecules of Venus are not. Thought by some scientists to be a metabolic product of microbes that live in the thick clouds that surround our sister planet, they would have to be considered as contemporary life.

by J. William Schopf and John Valley, of UCLA and the University of Wisconsin, Madison, started out with a long paragraph listing all possible abiotic sources for the Apex fossils since their discovery over 25 years ago.[3] A visit to the mineral collection at the National Museum of Natural History will make you sympathetic toward critics who support a geological origin of these fossils. The bizarre shapes of some of the collection's specimens, crystals slowly formed and extended into gravity-defying filaments, suggests that nature can sculpt some very strange-looking shapes, even those that look like filamentous bacteria. In the end, however, Schopf and Valley produced evidence that strongly supports the view that the Apex fossil microbe *Primaevifilum amoenum*, a long thin carbonaceous filament containing more than 40 individual cells, is a legitimate life-form.

As the technologies available to the paleobiologist continue to advance, the potential for ancient biomolecules to contribute to our understanding of the evolution of life, as well as the behavioral, physiological, ecological, and phylogenetic relationships of ancient life, will be limited only by our ability to find more fossils. They are out there; we just need to make ancient biomolecules a scientific priority. Some scientists, however, are not content to wait: a new field of research has already begun to make significant contributions to ancient biomolecule studies in the total absence of fossils and preserved biomolecules. Called ancient sequence reconstruction, it is entirely computer-based.

Paleogenetics

Most DNA has no function other than to store information. It doesn't do anything other than sit around and wait to be copied. In the absence of proteins, the polymerases, and other enzymes that make DNA functional, it is simply inactive. Unfortunately, as we have seen, ancient proteins don't have a great fossil record. The consensus is that they go back about 3 or 4 million years (although much longer if the work of Mary Schweitzer and her colleagues can be corroborated). Moreover, few ancient proteins are preserved and recovered intact; usually only a few short sequences can be determined. Of the 126 proteins identified in a 43,000-year-old mammoth, on average, a mere one-seventh of each protein's sequence was determined. While partial sequences may be of value in molecular systematics and phylogenetics, it is unrealistic to think that such tiny protein fragments would exhibit normal function or, for that matter, any function at all. And the older the ancient protein, the more degraded it is and the smaller the amount of sequence available for recovery. It's a trade-off. We can obtain ancient genes intact— complete genomes, in fact—but they are young, around a million years old. Or we can work with significantly older ancient proteins, but they are rarely recovered as a complete sequence. At present, more than a half billion years of ancient protein sequences are missing.

Luckily, ancient proteins can be obtained by means other than recovery of entire fossil sequences and, using that information, their synthesis in the laboratory. One alternative approach to the study of ancient proteins, called ancient sequence reconstruction or paleogenetics, involves experiments conducted solely via computer modeling.[4] Sequence reconstruction provides us with a prediction—an ancient sequence inferred from the comparison of sequences of related living organisms. For example, if a well-established phylogenetic tree is available, the sequences of a particular protein from each organism can be aligned on the tree. It helps if some of the sequences are from both living but primitive species and more recently derived or "modern" species. For example, if we were after an ancestral fish protein, sequences from a coelacanth (a living but primitive species) and the much more modern brown trout would be good choices. Trends in the patterns of mutations through time, and throughout the tree, are analyzed to create statistical algorithms, which predict the most likely ancestral sequences.

The technique is not amenable to all proteins; if a protein's sequence has too many substitutions over time, the sheer number of different possible combinations weakens the algorithm's ability to predict a realistic ancestral sequence. Alternatively, if the protein sequence is so stable as to have not undergone mutations for millions of years, there would be little to predict. In the abbreviated example provided here, the sequences of a theoretical five-residue protein from four different living mammals are provided in black, superimposed on their phylogenetic/evolutionary tree. By working backwards (right to left), selecting the residue statistically most likely to fill each position, it is possible to predict the most probable ancestral protein sequence, GDHHA, of the ancestor to all mammals. However, with apologies to the algorithms, *in silico*-generated sequences are only reasoned predictions—we cannot prove that a 15-million-year-old horse actually synthesized a predicted sequence. Obviously, our confidence in *in silico* predictions would be greatly enhanced by the availability of truly positive controls—deep time fossil-derived ancient sequences.

Great strides have been made in the application of ancient sequence reconstruction, a very active field of protein biochemistry. Its legitimacy and value have been established by, for example, the pharmaceutical industry's successful use of ancient sequence reconstruction in drug discovery. In one case, sequence reconstruction was used to understand why a popular cancer drug was effective against cancers caused by one type of oncogene but not others.[5] In studies of paleobiology and evolution,

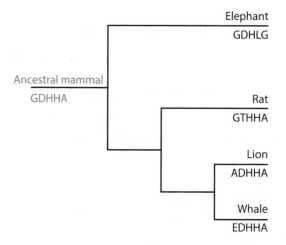

FIGURE 12.1. Prediction of the sequence of a five-residue ancestral protein by ancient sequence reconstruction. Selection of the residue most commonly used at each position predicts the ancestral sequence GDHHA. Prediction of the sequence ancestral to the group that contains lions, whales, and rats requires more information—a longer sequence. The chemists among you can predict the most likely sequence ancestral to the lion and whale.

ancient sequence reconstruction allows us to predict an ancient protein sequence, synthesize the corresponding DNA, produce the recombinant protein, and assay its function—all in the absence of a fossil. This approach has, for example, recently contributed to our understanding of evolution as it relates to brilliantly colored pigments, the structural molecules cellulose and lignin, and heat-stable enzymes—all topics that we have explored in prior chapters.

Although many reefs, especially those in the Caribbean, now face bleaching and death as a result of global warming, the reefs I encountered in Samoa were full of life and vibrant colors. Fluorescent pigments—which come in a wide variety of reds, blues, and greens—act as sun block for corals that live in shallow water, and they are thought to provide light for the photosynthetic activity of the symbiotic algae of the corals that live in deeper water where sunlight can't penetrate.

Mikhail Matz's laboratory at the University of Florida was interested in how the theoretical ancestral pigment—which is a protein—evolved to produce the wide array of fluorescent colors that are on display in living corals. To solve the problem, his team of scientists used the sequences of several related living species of coral that expressed various colors of fluorescent light to predict an ancestral sequence.[6] When they synthesized the ancestral fluorescent protein, they discovered that the most basal ancestor of these proteins, the very first of these fluorescent proteins to evolve, emitted green light.

In a Herculean effort, numerous additional intermediate ancestral sequences were predicted, and their corresponding proteins synthesized. The results showed that a total of 13 substitutions were required to produce a red fluorescent protein. The red fluorescent protein evolved from an ancestral green fluorescent protein, slowly, as numerous amino acid substitutions accumulated. Eventually, different ancient species of coral produced different ancestral intermediates that produced green, yellow-green, red-green, and red fluorescent colors. The publication that resulted from this work contained a famous photograph of a very colorful petri dish. Bacteria that synthesized recombinant versions of the differently colored fluorescent proteins were inoculated onto the surface of the soft growth medium in the petri dish, in the pattern of the phylogenetic tree that depicted the evolutionary divergence of the various pigments. Each branch of the tree was drawn with bacteria that expressed the appropriate ancestral fluorescent protein. After a short period of growth, the agar plate was exposed to ultraviolet light and lit up with all the colors of a tropical coral reef.

This newfound ability to "paint" with pigment-producing microbes has given rise to an annual competition sponsored by the American Society for Microbiology, the Agar Art Contest. Winning entries are exhibited at the Agar Art Gallery at the Society's meetings. A recent winner, Andrea Héjja from Budapest, produced "Mondays like these . . . ," a recreation of Edvard Munch's *The Scream*. She was awarded one hundred dollars and a T-shirt, a far cry from the $120 million that *The Scream* sold for in 2012. Nevertheless, the competition gives short shrift to the ugly rumor that science and art don't mix.

The most ambitious use of ancestral sequence reconstruction may be recent attempts to determine if characteristics of ancient proteins can inform us about the environment in which they originated. In other words, can ancient proteins tell us exactly where on Earth life evolved? One popular theory suggests that the ample supplies of carbon dioxide, water, hydrogen, and minerals available at hydrothermal vents located deep beneath the water's surface along the mid-oceanic ridges may have provided the right mix of conditions to create life. If ancient life evolved within the thermal gradients present near deep-sea hydrothermal vents (some reaching temperatures of 750 degrees Fahrenheit), then the proteins of those life-forms must have been able to survive and function at very high temperatures. If ancestral sequence reconstruction could produce examples of really deep time proteins that functioned best at very high temperatures, it would provide at least circumstantial evidence that the very first proteins, and the organisms of which they were a part, existed in an environment characterized by hot temperatures.

To study this possibility, Akihiko Yamagishi and his colleagues at the University of Tokyo picked an enzyme that helps to make biomolecules essential to all known life-forms, the small subunits that are precursors of both RNA and DNA.[7]

They then built a large phylogenetic tree that contained representatives of all the various types of living organisms, including "primitive" living bacteria, the Archaea—organisms that superficially look like bacteria but are more closely related to more advanced life—and plants and animals. The tree took them back 4 billion years, yet the enzyme's DNA sequences remained mostly unchanged. Given its critical function, evolution hasn't fiddled with it terribly; mutations, when they occurred, were most often lethal and so not passed to subsequent generations. Nevertheless, numerous different versions of the 4-billion-year-old ancestral protein were predicted. Recombinant forms of these ancestral proteins were made and then tested at various temperatures.

Yamagishi found that their activity, unlike that of their counterparts in most living organisms, increased as temperatures rose. The optimal temperature for the ancient enzymes was about 176 degrees Fahrenheit (for comparison, at 108 degrees Fahrenheit, the human brain begins to suffer irreparable damage). Amazingly, many of the ancestral proteins were active even at boiling temperatures. Yamagishi and his colleagues concluded that the ancestral enzymes could function quite well in a very hot environment, as could their parent organisms—additional evidence that supports deep-sea hydrothermal vents as the cradle of life.

Phylogenetic trees that span billions of years will inevitably include many organisms and huge numbers of ancestral proteins. Proteins at adjacent forks or nodes* of the tree may, or may not,

---

* There are three nodes (marked with asterisks) in the diagram displayed earlier in this chapter. The most likely sequence at the node or fork with branches to the whale and lion should read ADHHA. G (in the rat and elephant) is very similar structurally and chemically to A, whereas E is very different and represents a radical change.

have different sequences. However, the greater the time and the number of nodes that separate the various organisms and their proteins, the greater the chance for sequence differences.

Eric Gaucher and his colleagues at the University of Florida, Gainesville, studied a different set of proteins essential to life, factors that aid in the elongation of proteins as they are synthesized. They used ancient sequence reconstruction to predict sequences of ancestral elongation factor proteins over a period of 3 billion years, from 3.5 to half a billion years ago.[8] Their results agreed with and expanded upon those obtained by Yamagishi. In addition to showing that the original 3-billion-year-old protein was thermophilic, his team then went on to predict the sequences of the proteins at numerous nodes during the course of 3 billion years of evolution. Over the course of those 3 billion years, the optimal temperature of the predicted ancient elongation factors' activities decreased, gradually, by over 50 degrees Fahrenheit; the factor became less thermophilic. This supports the idea that as organisms diversified and moved away from the hot vents at the ocean's floor to cooler environments, they no longer required their proteins to operate at exceedingly high temperatures. Mutations that lowered the enzymes' optimal temperatures were thus tolerated and passed on.

Commercialization of Ancient Biomolecules

It is estimated that a half teraton of plant biomass, much of which is wood, exists on the earth's surface. About a quarter of that is the heterogeneous polymeric biomolecule lignin, which, after cellulose and chitin, is the most abundant biomolecule on earth. The construction industry takes advantage of the size and rigidity of wood, a direct result of the structural support provided by lignin, but as a bioresource, we do little with lignin other than burn it. But if we stop burning lignin and instead think of it as a renewable bioresource, at some point in the future, it could replace oil as the source of the tens of thousands of commercial chemicals that support the existence of modern civilization. Replace oil? Actually, it is not so far-fetched. Yulin Deng and his colleagues at the Georgia Institute of Technology are developing processes that can do just that. Our old friend pyrolysis can be used to produce an intermediate bio-oil from lignin, and Deng's lab has invented a process that, at relatively low temperatures, produces commercial-grade hydrocarbons from that oil.

In nature, the removal of lignin from wood allows cellulose-digesting microbes access to cellulose fibrils and speeds the decomposition of dead plants. However, few organisms can degrade lignin. One exception is fungi that make a lignin-degrading (or lignolytic) enzyme. Interested in the function and evolution of the lignolytic enzyme, Iván Fernández and his mentors at the Complutense University of Madrid used ancestral sequence reconstruction to predict the sequences of the enzyme over a period of several hundred million years.[9] With the predicted ancestral sequences in hand, they synthesized the corresponding genes that encoded the enzymes and then made recombinant versions of them. They found that the ancestral protein sequences that were predicted to have existed before and after about 200 million years ago had different substrate preferences. The older enzymes were better at degrading lignin from gymnosperms (for example, conifer trees like pines, spruces, and redwoods), while the younger enzymes were more efficient at degrading lignin from angiosperms (for example, oaks, maples, and apple trees). Is it simply coincidental that molecular clock-based estimates of the appearance of the very earliest angiosperms give an age of about 200 million years? Flowering plants, the angiosperms, now dominate the planet's fauna with about 300,000 different species, compared to the thousand or so species of gymnosperms, most of which are conifers. As nature transitioned from conifer-dominated forests to angiosperm-dominated forests, it appears that it also took care to provide for the natural decomposition of the new forests.

In addition to advancing our understanding of ancient life, this work has broad practical applications. Numerous patents have already been granted for ancestral enzymes that can be used commercially, on an industrial scale, operating at higher temperatures so as to increase output and reduce costs. One can easily imagine a thermophilic lignolytic enzyme based on an ancestral protein that is more efficient in the industrial-scale processing of lignin. And Fernández's work suggests that an ancestral enzyme that is both thermophilic and optimized to degrade conifer lignin would be of great commercial value, given that most commercial tree farms produce rapidly growing conifer trees.

The Ultimate Ancient Biomolecule

In today's world, far removed from the collecting expeditions of Alexander von Humboldt, Charles Darwin, and Thomas Huxley in the early and mid-nineteenth century, every biologist dreams of discovering a new species, perhaps in the mountains of Papua New Guinea or miles deep in the South Pacific Ocean. And many do, as thousands of new living species are described each year, many of them insects. On very rare but happy occasions, species thought to be extinct are rediscovered in the wild. One recent example of these so-called Lazarus species is the Vietnamese mouse-deer, the world's smallest hoofed animal.[10]

In an extremely rare case, scientists discovered not only an extinct species, but an extinct genus, *Metasequoia*. Prior to its rediscovery, the genus consisted of five different extinct species of redwood trees, with a fossil record that goes back well over a hundred million years (the California redwood is a different genus, *Sequoia*). The leaves of one of these species, *Metasequoia occidentalis*, are some of the more common fossils to be found at the Fossil Bowl in Clarkia, Idaho; cones and seeds are also preserved in the Miocene lake sediments there. The fossil record of another species, *Metasequoia glyptostroboides,* the dawn redwood, extends back about 50 million years to the Eocene. In the 1940s, the tree was found growing near Moudao in central China.[11]

There are, of course, large numbers—billions of years' worth—of truly extinct species. If we can't find living specimens of these long-extinct species, can we resurrect them from their fossils? While this may seem like science fiction, ancient species have, in fact, already been resurrected. Two were recovered from permafrost of the arctic tundra in Siberia. The first is *Silene stenophylla*, a small flowering plant sometimes called the narrow-leafed campion. Approximately 32,000 years ago, rodents foraged for plants, brought them back to their burrows, and stored them for later consumption; some burrows contained thousands of *Silene stenophylla* seeds. Fortunately, many of the burrows were constructed within existing permafrost. The burrows subsequently collapsed, and the seeds were entombed. The covering sediments quickly froze, and the seeds were eventually buried under more than a hundred feet of additional permafrost, some of which was as much as 80 percent ice.

Discovered by a team of scientists led by Svetlana Yashina of the Russian Academy of Sciences, numerous seedpods of *Silene stenophylla* were recovered. Some were ripe and open, others closed and immature.[12] Initial attempts to grow the plant from the 32,000-year-old seeds were unsuccessful, so Yashina tried to cultivate tissue dissected from the inner portion of the seedpod in test tubes containing growth medium. It worked; individual cells had remained viable for over 300 centuries. Tiny shoots appeared and eventually grew into rooted plants. With time, and cultivation, the plants produced flowers and fertile viable seeds. The scientists were very conservative and assigned the resurrected plant to a living species. But its morphology differed significantly from the *Silene stenophylla* that grows today in the mountains of Japan and the arctic tundra of northeastern Siberia. The ancient plant produced twice as many buds, and its petals were narrower and terminated in two lobes, rather than several, as in its modern relative. Perhaps, at some point, a molecular analysis of the permafrost-derived plant will determine whether or not the plant may in fact be a new species.

The cells of *Silene stenophylla* are an exceptional example of intact cells retaining their viability after being frozen for tens of thousands of years. It almost never happens. Scientists in the biotechnology industry routinely freeze cells for subsequent recovery; the umbilical cord blood and egg and sperm bank businesses are built on that technology. But the protocols used to freeze cells for preservation are extremely demanding. Cells must be frozen in a well-defined solution of chemicals, at a slow and tightly controlled rate of cooling. If they are frozen too quickly by placing a vial of cells directly into liquid nitrogen, they will die. If they are frozen at a relatively warm temperature, by throwing them into a snowbank, they will die. If they are subjected to repeated warming and cooling, as they would sitting next to the ice cream in your home's freezer, they will die.

In a television documentary on the cloning of mammoths a decade or so ago, mammoth tissue from Siberian permafrost was shown being processed by scientists dressed in white gowns and blue masks at what was obviously a state-of-the-art laboratory. During the final few seconds of the film, just before the credits rolled down the screen, the camera zoomed in on a small petri dish and a microscopic view of "cells," as the narrator hinted at the isolation of viable mammoth cells. Ultimately, however, living cells were not isolated. Nearly a decade later, however, Akira Iritani published a manuscript in which his team reported the isolation of numerous nucleus-like structures from a 28,000-year-old Siberian mammoth.[13] The structures were shown to contain DNA-binding proteins that are found only in nuclei. The researchers even injected the mammoth cell "nuclei" into mouse eggs in an attempt to make an embryo. The eggs died.

A second and very different kind of resurrection produced an organism that has no known living counterpart. During their lifelong studies of viruses, the husband-and-wife team of Jean-Michel Claverie and Chantal Abergel at the Aix-Marseille Université in France didn't discover a new species—they discovered an entirely new order, Megavirales, a unique and surprisingly weird group of viruses. In fact, there has been discussion as to whether or not Megavirales represents a new domain.* Viruses (with the exception of Megavirales) are fairly simple to define. They are very small, and they can't make their own proteins; they must enter their host and hijack its protein-synthesis machinery for its own purposes. They don't reproduce by dividing; rather, they simply manufacture large numbers of clones of themselves, kill the host, or not, and escape to find another.

Claverie and Abergel travelled the world, from Chile to Australia and Siberia, in search of different types of viruses. Their

---

* There are currently three domains of living organisms: Bacteria, Archaea— these two differ in the lipid composition of their membranes—and Eukarya, in which cells contain a nucleus.

discovery of the first representative of Megavirales, *Mimivirus,* occurred in 2003.[14] *Mimivirus* and several other related viruses subsequently discovered by the two scientists are huge, about 15 times as large as the coronavirus SARS-CoV-2 that causes COVID-19. And while most viruses contain a minimal number of genes, usually just those needed to take over the host and encode their own proteins, the Megavirales have huge genomes, encoding more proteins than are present in some bacteria. In another member of the group, *Pandoravirus,* only a tiny percentage of its genome, which encodes as many as 2,500 proteins, is related to known DNA sequences; most of its proteins are unique—unrelated to any known protein on Earth; no one has a clue as to their function.

When Claverie and Abergel read about the resurrection of *Silene stenophylla* from Siberian permafrost, they wondered if the permafrost contained Megavirales. They devised an experiment based on the observation that Megavirales viruses infect only protozoans of the genus *Amoeba.*[15] They cultured healthy amoebae and added small amounts of permafrost soil from various locations. Sure enough, some of the amoebae began to die. When examined, a new type of these giant complex viruses was found inside the dying protozoans. The new virus, *Pithovirus sibericum,* is also huge, 1.5 microns in length. The ancient virus has not been found at any other place on Earth; it appears to be an organism that had been truly extinct, perhaps an evolutionary intermediate in the evolution of the modern giant viruses.

Better Living through Chemistry

Craig Venter, as a scientist at the National Institute of Neurological Disorders and Stroke in Bethesda, Maryland, popularized the use of a technique called shotgun sequencing, which was critical to the successful sequencing of the human genome.

Subsequently, Venter took the radical new technology and helped to found the very first genomics-based biotechnology companies, Human Genome Sciences in 1992 and Celera Genomics in 1998. Venter's equity in those companies provided the funds that allowed him to create the Venter Institute, which, unlike product-driven Human Genome Sciences, focused much of its energies on basic research. One of Venter's quests was to determine the minimum number of genes required to make a fully viable, self-replicating life-form. Many common bacteria, like *Escherichia coli*, have well over 4,000 genes, so for their experimental organism the researchers picked *Mycoplasmum genitalium*, with its much smaller genome of only 525 genes.

They began by chemically synthesizing all 525 genes and assembling them into one contiguous circular structure, essentially a synthetic chromosome. Next, they replaced the natural chromosome of the bacterium with the synthetic one to create living bacteria that they could grow in the laboratory. They then started to remove genes one by one. If removal of a gene was fatal to the bacterium, it was noted, the gene was returned, and a different gene was deleted; one by one, every gene was tested. The final product, a fully viable bacterium called JCVI Syn3.0, contained only 473 genes.[16] The progeny of their creation was a totally synthetic life-form.

Given what we have learned so far, what are the possibilities if we want to bring back long-extinct creatures? Is it realistic to think we can find a frozen, yet viable, mammoth cell? We most certainly will never find frozen tissue of a Tasmanian tiger. And the odds of finding a fully functional nucleus in anything other than a viable cell are zero. If we have its entire genome, can we make a mammoth from scratch, like Venter and his colleagues did with JCVI Syn3.0? Keep in mind that the genome of an elephant is nearly 7,000 times larger than that of JCVI Syn3.0. And to make a viable mammoth, we will need its entire genome, not just the minuscule (less than 2 percent) portion of it that encodes proteins. Many genes (perhaps most genes) do not function to encode proteins; rather, they act as regulators that dictate where and when other genes give rise to proteins. Of the 473 genes in JCVI Syn3.0, 149 had no known function; however, every one of them was essential for the organism's viability.

None of these potential options for the resurrection of extinct species—de-extinction, as it is sometimes called—are currently viable. However, George Church of Harvard University isn't convinced that cloning, replacement, or synthesis of a complete ancient genome is needed, or even desirable.[17] Perhaps, as Church has suggested, only a few genes need to be changed to resurrect an ancient species. After all, we have all heard that chimpanzees share 99 percent of our DNA. How hard could it be?

First, it is worth noting that the 99 percent figure is terribly misleading. Given that the average mammalian gene is 10,000 or more subunits (nucleotides) long, a difference of just 1 percent could mean a hundred substitutions per gene, for every gene. More than enough to ensure that every chimp protein is significantly different from its homologous human protein—or that every mammoth protein is different from every elephant protein. Moreover, even if we could create an African elephant with long red hair, smaller ears, and a new hemoglobin gene, would it really be a mammoth, or would it be a Barnum and Bailey oddity? Would not the millions of dollars required to "de-extinct" an animal destined for a zoo be better used to prevent extinction of the many endangered organisms that live today?

The concept of de-extinction itself is as contentious as the science of resurrection is difficult. While there exist living cells that have an ancient gene as part of their genome, albeit in a research laboratory or for commercial production of a thermophilic enzyme, the ultimate ancient biomolecule, or more accurately, the ultimate assemblage of ancient biomolecules—a de-extincted organism—exists only in our dreams.

Although ancient biomolecule research is in its infancy, its potential applications are manifest. Taxonomy and the history of evolution are areas where ancient biomolecules will make immediate, numerous, and important contributions. Information gleaned from studies of ancient biomolecules will also inform us as to the physiology and behavior of extinct organisms. Even paleoenvironmental parameters—such as confirming the temperature of the earth 55 million years ago—will be defined through proxies provided by ancient biomolecules. And even with all that, we may be just scratching the surface.

Paleobiology is an active, dynamic, and expanding field, encompassing everything from the discovery of the permineralized bones of new extinct species to the prediction of the sequences and functions of life's first proteins. When Alan Grant, the dinosaur expert in *Jurassic Park*, first saw a living dinosaur, a towering brachiosaur, one of the first things he said was "it's a warm-blooded creature!" Even if we never resurrect the brachiosaur, today's young students, and the scientists they will become, may be able to reach similarly startling conclusions about ancient life.

NOTES

Chapter 1. A Blood-Engorged Mosquito

1. Engel, M. S., and Grimaldi, D. A., 2004. New light shed on the oldest insect. *Nature*, 427(6975), pp. 627–630.

2. Pape, T., Blagoderov, V., and Mostovski, M. B., 2011. Order Diptera Linnaeus, 1758. In Zhang, Z.-Q. (ed.), Animal biodiversity: An outline of higher-level classification and survey of taxonomic richness. *Zootaxa*, 3148(1), pp. 222–229.

3. McPhee, J., 1981. *Basin and range*. Macmillan, p. 8.

4. Pape, T., Blagoderov, V., and Mostovski, M. B., 2011. Order Diptera Linnaeus, 1758. In Zhang, Z.-Q. (ed.), Animal biodiversity: An outline of higher-level classification and survey of taxonomic richness. *Zootaxa*, 3148(1), pp. 222–229.

5. Sánchez-Baracaldo, P., and Cardona, T., 2020. On the origin of oxygenic photosynthesis and Cyanobacteria. *New Phytologist*, 225(4), pp. 1440–1446.

6. Harbach, R. E., and Greenwalt, D., 2012. Two Eocene species of Culiseta (Diptera: Culicidae) from the Kishenehn Formation in Montana. *Zootaxa*, 3530(1), pp. 25–34.

7. Borkent, A., and Grimaldi, D. A., 2004. The earliest fossil mosquito (Diptera: Culicidae), in mid-Cretaceous Burmese amber. *Annals of the Entomological Society of America*, 97(5), pp. 882–888; Borkent, A., and Grimaldi, D. A., 2016. The Cretaceous fossil *Burmaculex antiquus* confirmed as the earliest known lineage of mosquitoes (Diptera: Culicidae). *Zootaxa*, 4079(4), pp. 457–466.

8. Izmailova, D. Z., Serebrennikov, V. M., Mozzhelina, T. K., and Serebrennikova, O. V., 1996. Features of the molecular composition of metalloporphyrins of crude oils of the Volga-Urals oil- and gas-bearing province. *Petroleum Chemistry*, 36, 111–117; Schweitzer, M. H., Marshall, M., Carron, K., Bohle, D. S., Busse, S. C., Arnold, E. V., Barnard, D., Horner, J. R., and Starkey, J. R., 1997. Heme compounds in dinosaur trabecular bone. *Proceedings of the National Academy of Sciences*, 94, 6291–6296.

9. Grimes, S. T., Brock, F., Rickard, D., Davies, K. L., Edwards, D., Briggs, D. E., and Parkes, R. J., 2001. Understanding fossilization: Experimental pyritization of plants. *Geology*, 29(2), pp. 123–126.

10. Briggs, D. E., 2013. A mosquito's last supper reminds us not to underestimate the fossil record. *Proceedings of the National Academy of Sciences, 110*(46), pp. 18353–18354.

11. Greenwalt, D. E., Goreva, Y. S., Siljeström, S. M., Rose, T., and Harbach, R. E., 2013. Hemoglobin-derived porphyrins preserved in a Middle Eocene blood-engorged mosquito. *Proceedings of the National Academy of Sciences, 110*(46), pp. 18496–18500.

12. Greenwalt, D. E., Rose, T. R., Siljestrom, S. M., Goreva, Y. S., Constenius, K. N., and Wingerath, J. G., 2014. Taphonomy of the fossil insects of the middle Eocene Kishenehn Formation. *Acta Palaeontologica Polonica, 60*(4), pp. 931–947.

Chapter 2. In Situ

1. Reviewed in Whitehouse, M. J., Dunkley, D. J., Kusiak, M. A., and Wilde, S. A., 2019. On the true antiquity of Eoarchean chemofossils—Assessing the claim for Earth's oldest biogenic graphite in the Saglek Block of Labrador. *Precambrian Research, 323*, pp. 70–81.

2. Nutman, A. P., Bennett, V. C., Friend, C. R., Van Kranendonk, M. J., and Chivas, A. R., 2016. Rapid emergence of life shown by discovery of 3,700-million-year-old microbial structures. *Nature, 537*(7621), pp. 535–538.

3. El Albani, A., Bengtson, S., Canfield, D. E., Riboulleau, A., Bard, C. R., Macchiarelli, R., Pemba, L. N., Hammarlund, E., Meunier, A., Mouélé, I. M., Benzerara, K., Bernard, S., Boulvais, P., Chaussidon, M., Cesari, C., Fontaine, C., Chi-Fru, E., Ruiz, J.M.G., Gauthier-Lafaye, F., Mazurier, A., Pierson-Wickmann, A. C., Rouxel, O., Trentesaux, A., Vecoli, M., Versteegh, G.J.M., White, L., Whitehouse, M., and Bekker, A., 2014. The 2.1 Ga old Francevillian biota: Biogenicity, taphonomy and biodiversity. *PLoS One, 9*(6), e99438.

4. Han, T. M., and Runnegar, B., 1992. Megascopic eukaryotic algae from the 2.1-billion-year-old Negaunee Iron-Formation, Michigan. *Science, 257*(5067), pp. 232–235; Butterfield, N. J., 2000. *Bangiomorpha pubescens* n. gen., n. sp.: Implications for the evolution of sex, multicellularity, and the Mesoproterozoic/Neoproterozoic radiation of eukaryotes. *Paleobiology, 26*(3), pp. 386–404.

5. Zhu, S., Zhu, M., Knoll, A. H., Yin, Z., Zhao, F., Sun, S., Qu, Y., Shi, M., and Liu, H., 2016. Decimetre-scale multicellular eukaryotes from the 1.56-billion-year-old Gaoyuzhuang Formation in North China. *Nature Communications, 7*(1), pp. 1–8.

6. Horodyski, R. J., 1982. Problematic bedding-plane markings from the middle Proterozoic Appekunny argillite, Belt Supergroup, northwestern Montana. *Journal of Paleontology, 56*, pp. 882–889.

7. Retallack, G. J., Dunn, K. L., and Saxby, J., 2013. Problematic Mesoproterozoic fossil *Horodyskia* from Glacier National Park, Montana, USA. *Precambrian Research, 226*, pp. 125–142.

8. Rule, R. G., and Pratt, B. R., 2019. The pseudofossil Horodyskia: Flocs and flakes on microbial mats in a shallow Mesoproterozoic sea (Appekunny Formation, Belt Supergroup, western North America). *Precambrian Research, 333*, p. 105439.

9. Brasier, M., and Antcliffe, J., 2004. Decoding the Ediacaran enigma. *Science,* 305(5687), pp. 1115–1117.

10. Schiffbauer, J. D., Selly, T., Jacquet, S. M., Merz, R. A., Nelson, L. L., Strange, M. A., Cai, Y., and Smith, E. F., 2020. Discovery of bilaterian-type through-guts in cloudinomorphs from the terminal Ediacaran Period. *Nature Communications, 11*(1), pp. 1–12.

11. Love, G. D., Grosjean, E., Stalvies, C., Fike, D. A., Grotzinger, J. P., Bradley, A. S., Kelly, A. E., Bhatia, M., Meredith, W., Snape, C. E., and Bowring, S.A., 2009. Fossil steroids record the appearance of Demospongiae during the Cryogenian period. *Nature, 457*(7230), pp. 718–721.

12. Turner, E. C., 2021. Possible poriferan body fossils in early Neoproterozoic microbial reefs. *Nature, 596*(7870), pp. 87–91.

13. Vellekoop, J., Sluijs, A., Smit, J., Schouten, S., Weijers, J. W., Damsté, J.S.S., and Brinkhuis, H., 2014. Rapid short-term cooling following the Chicxulub impact at the Cretaceous–Paleogene boundary. *Proceedings of the National Academy of Sciences, 111*(21), pp. 7537–7541.

14. Schaefer, B., Grice, K., Coolen, M. J., Summons, R. E., Cui, X., Bauersachs, T., Schwark, L., Böttcher, M. E., Bralower, T. J., Lyons, S. L., and Freeman, K. H., 2020. Microbial life in the nascent Chicxulub crater. *Geology, 48*(4), pp. 328–332.

15. Bobrovskiy, I., Hope, J. M., Ivantsov, A., Nettersheim, B. J., Hallmann, C., and Brocks, J. J., 2018. Ancient steroids establish the Ediacaran fossil *Dickinsonia* as one of the earliest animals. *Science, 361*(6408), pp. 1246–1249.

16. Melendez, I., Grice, K., and Schwark, L., 2013. Exceptional preservation of Palaeozoic steroids in a diagenetic continuum. *Scientific Reports, 3*, p. 2768.

17. Gould, S. J., 1990. *Wonderful Life: The Burgess Shale and the Nature of History.* W. W. Norton & Company.

18. Shu, D.-G., Conway Morris, S., Han, J., Zhang, Z.-F., Yasui, K., Janvier, P., Chen, L., Zhang, X.-L., Liu, J.-N., Li, Y., and Liu, H.-Q., 2003. Head and backbone of the Early Cambrian vertebrate *Haikouichthys. Nature, 421*(6922), pp. 526–529.

19. Stawski, D., Rabiej, S., Herczynska, L., and Draczynski, J., 2008. Thermogravi-metric analysis of chitins of different origin. *J. Therm. Anal. Calorim., 93*, 489–494, 10.1007/s10973-007-8691-6.

20. Ehrlich, H., Rigby, J. K., Botting, J. P., Tsurkan, M. V., Werner, C., Schwille, P., Petrášek, Z., Pisera, A., Simon, P., Sivkov, V. N., Vyalikh, D. V., Molodtsov, S. L., Kurek, D., Kammer, M., Hunoldt, S., Born, R., Stawski, D., Steinhof, A., Bazhenov, V. V., and Geisler, T., 2013. Discovery of 505-million-year-old chitin in the basal demosponge *Vauxia gracilenta. Scientific Reports, 3*(1), pp. 1–6.

21. Cody, G. D., Gupta, N. S., Briggs, D.E.G., Kilcoyne, A.L.D., Summons, R. E., Kenig, F., Plotnick, R. E., and Scott., A. C., 2011. Molecular signature of chitin-protein complex in Paleozoic arthropods. *Geology*, 39, pp. 255–258; Wysokowski, M., Zatoń, M., Bazhenov, V. V., Behm, T., Ehrlich, A., Stelling, A. L., Hog, M., and Ehrlich, H., 2014. Identification of chitin in 200-million-year-old gastropod egg capsules. *Paleobiology*, 40(4), pp. 529–540; Weaver, P. G., Doguzhaeva, L. A., Lawver, D. R., Tacker, R. C., Ciampaglio, C. N., Crate, J. M., and Zheng, W., 2011. Characterization of organics consistent with β-chitin preserved in the Late Eocene cuttlefish *Mississaepia mississippiensis*. *PLoS One*, 6(11), p. e28195; Stankiewicz, B. A., Briggs, D. E., Evershed, R. P., Flannery, M. B., and Wuttke, M., 1997. Preservation of chitin in 25-million-year-old fossils. *Science*, 276(5318), pp. 1541–1543.

22. Cody, G. D., Gupta, N. S., Briggs, D.E.G., Kilcoyne, A.L.D., Summons, R. E., Kenig, F., Plotnick, R. E., and Scott., A. C., 2011. Molecular signature of chitin-protein complex in Paleozoic arthropods. *Geology*, 39, pp. 255–258.

23. Rigby, J. K. Sponges of the Burgess Shale (Middle Cambrian), British Columbia, 1986. *Palaeontographica Canadiana*, 2, pp. 1–105.

24. Butterfield, N. J., 1990. Organic preservation of non-mineralizing organisms and the taphonomy of the Burgess Shale. *Paleobiology*, pp. 272–286.

Chapter 3. The Purple Fossil

1. Fleming, J. F., Kristensen, R. M., Sørensen, M. V., Park, T.Y.S., Arakawa, K., Blaxter, M., Rebecchi, L., Guidetti, R., Williams, T. A., Roberts, N. W., and Vinther, J., 2018. Molecular palaeontology illuminates the evolution of ecdysozoan vision. *Proceedings of the Royal Society B*, 285(1892), p. 20182180.

2. Parker, A. R., 2000. 515 million years of structural colour. *Journal of Optics A: Pure and Applied Optics*, 2(6), p. R15.

3. Erwin, D. H., 2015. *Extinction: How Life on Earth Nearly Ended 250 Million Years Ago—Updated Edition*. Princeton University Press.

4. Johnson, A., and White, N. D., 2014. Ocean acidification: The other climate change issue. *American Scientist*, 102(1), pp. 60–63.

5. Blumer, M., 1951. Fossile Kohlenwasserstoffe und Farbstoffe in Kalksteinen (Geochemische Untersuchungen III.). *Mikrochemie vereinigt mit Mikrochimica acta*, 36(2), pp. 1048–1055.

6. Wolkenstein, K., 2015. Persistent and widespread occurrence of bioactive quinone pigments during post-Paleozoic crinoid diversification. *Proceedings of the National Academy of Sciences*, 112(9), pp. 2794–2799.

7. Kubin, A., Wierrani, F., Burner, U., Alth, G., and Grunberger, W., 2005. Hypericin—the facts about a controversial agent. *Current Pharmaceutical Design*, 11(2), pp. 233–253.

8. O'Malley, C. E., Ausich, W. I., and Chin, Y. P., 2016. Deep echinoderm phylogeny preserved in organic molecules from Paleozoic fossils. *Geology, 44*(5), pp. 379–382.

9. Wolkenstein, K., Gross, J. H., and Falk, H., 2010. Boron-containing organic pigments from a Jurassic red alga. *Proceedings of the National Academy of Sciences, 107*(45), pp. 19374–19378.

10. Pidot, S., Ishida, K., Cyrulies, M., and Hertweck, C., 2014. Discovery of clostrubin, an exceptional polyphenolic polyketide antibiotic from a strictly anaerobic bacterium. *Angewandte Chemie, 126*(30), pp. 7990–7993.

11. Barden, H. E., Behnsen, J., Bergmann, U., Leng, M. J., Manning, P. L., Withers, P. J., Wogelius, R. A., and Van Dongen, B. E., 2015. Geochemical evidence of the seasonality, affinity and pigmenation of *Solenopora jurassica*. *PLoS One, 10*(9), p. e0138305.

12. Vogt, P. R., Eshelman, R. E., and Godfrey, S. J., 2018. Calvert Cliffs: Eroding mural escarpment, fossil dispensary, and paleoenvironmental archive in space and time. *Smithsonian Contributions to Paleobiology, 100*, pp. 3–44.

13. Nance, J. R., Armstrong, J. T., Cody, G. D., Fogel, M. L., and Hazen, R. M., 2015. Preserved macroscopic polymeric sheets of shell-binding protein in the Middle Miocene (8 to 18 Ma) gastropod Ecphora. *Geochemical Perspectives Letters, 1*, pp. 1–9.

14. Johnson, K. R., Miller, I. M., and Pigati, J. S., 2014. The snowmastodon project. *Quaternary Research, 82*(3), pp. 473–476.

15. Weigelt, J., 1935. Lophiodon in der oberen Kohle des Geiseltales. *Nova Acta Leopoldina N. F., 3*(14), pp. 369–402.

16. Rember, W. C., 1991. Stratigraphy and paleobotany of Miocene lake sediments near Clarkia, Idaho (unpublished doctoral dissertation, University of Idaho).

17. Weigelt, J., and Noack, K., 1931. Über Reste von Blattfarbstoffen in Blättern aus der Geiseltal-Braunkohle Mitteleocän. *Nova Acta Leopoldina, N. F. 1*, 87.

18. Niklas, K. J., and Chaloner, W. G., 1976. Chemotaxonomy of some problematic Palaeozoic plants. *Review of Palaeobotany and Palynology, 22*(2), pp. 81–104; Henderson, W., Wollrab, V., and Eglinton, G., 1968. Identification of steroids and triterpenes from a geological source by capillary gas–liquid chromatography and mass spectrometry. *Chemical Communications (London)*, (13), pp. 710–712.

19. Niklas, K. J., and Giannasi, D. E., 1977. Flavonoids and other chemical constituents of fossil Miocene Zelkova (Ulmaceae). *Science, 196*(4292), pp. 877–878.

20. McNamara, M. E., Orr, P. J., Kearns, S. L., Alcalá, L., Anadón, P., and Peñalver, E., 2016. Reconstructing carotenoid-based and structural coloration in fossil skin. *Current Biology, 26*(8), pp. 1075–1082.

Chapter 4. The Black Pigment

1. Brenner, M., and Hearing, V. J., 2008. The protective role of melanin against UV damage in human skin. *Photochemistry and Photobiology, 84*(3), pp. 539–549.

2. McNamara, M. E., Rossi, V., Slater, T. S., Rogers, C. S., Ducrest, A. L., Dubey, S., and Roulin, A., 2021. Decoding the evolution of melanin in vertebrates. *Trends in Ecology & Evolution, 36*(5), pp. 430–443.

3. Lucking, R., Huhndorf, S., Pfister, D. H., Plata, E. R., and Lumbsch, H. T., 2009. Fungi evolved right on track. *Mycologia, 101*(6), pp. 810–822; and see Liu, R., Xu, C., Zhang, Q., Wang, S., and Fang, W., 2017. Evolution of the chitin synthase gene family correlates with fungal morphogenesis and adaption to ecological niches. *Scientific Reports, 7*, p. 44527.

4. McNamara, M. E., Rossi, V., Slater, T. S., Rogers, C. S., Ducrest, A. L., Dubey, S., and Roulin, A., 2021. Decoding the evolution of melanin in vertebrates. *Trends in Ecology & Evolution, 36*(5), pp. 430–443.

5. Glass, K., Ito, S., Wilby, P. R., Sota, T., Nakamura, A., Bowers, C. R., Vinther, J., Dutta, S., Summons, R., Briggs, D. E., and Wakamatsu, K., 2012. Direct chemical evidence for eumelanin pigment from the Jurassic period. *Proceedings of the National Academy of Sciences, 109*(26), pp. 10218–10223.

6. Joyce, W. G., Micklich, N., Schaal, S. F., and Scheyer, T. M., 2012. Caught in the act: The first record of copulating fossil vertebrates. *Biology Letters, 8*(5), pp. 846–848.

7. Carpenter, K., 1998. Evidence of predatory behavior by carnivorous dinosaurs. *Gaia, 15*, pp. 135–144.

8. Lindgren, J., Sjövall, P., Carney, R. M., Uvdal, P., Gren, J. A., Dyke, G., Schultz, B. P., Shawkey, M. D., Barnes, K. R., and Polcyn, M. J., 2014. Skin pigmentation provides evidence of convergent melanism in extinct marine reptiles. *Nature, 506*(7489), pp. 484–488.

9. McNamara, M. E., Orr, P. J., Kearns, S. L., Alcalá, L., Anadón, P., and Peñalver, E., 2016. Reconstructing carotenoid-based and structural coloration in fossil skin. *Current Biology, 26*(8), pp. 1075–1082.

10. Labandeira, C. C., Yang, Q., Santiago-Blay, J. A., Hotton, C. L., Monteiro, A., Wang, Y. J., Goreva, Y., Shih, C., Siljeström, S., Rose, T. R., and Dilcher, D. L., 2016. The evolutionary convergence of mid-Mesozoic lacewings and Cenozoic butterflies. *Proceedings of the Royal Society B: Biological Sciences, 283*(1824), p. 20152893.

11. Parker, A.R., 2000. 515 million years of structural colour. *Journal of Optics A: Pure and Applied Optics, 2*(6), p. R15.

12. Fleming, J. F., Kristensen, R. M., Sørensen, M. V., Park, T.Y.S., Arakawa, K., Blaxter, M., Rebecchi, L., Guidetti, R., Williams, T. A., Roberts, N. W., and Vinther, J., 2018. Molecular palaeontology illuminates the evolution of ecdysozoan vision. *Proceedings of the Royal Society B, 285*(1892), p. 20182180.

13. Ghosh, D. D., Lee, D., Jin, X., Horvitz, H. R., and Nitabach, M. N., 2021. *C. elegans* discriminates colors to guide foraging. *Science, 371*(6533), pp. 1059–1063.

14. Tanaka, G., Parker, A. R., Hasegawa, Y., Siveter, D. J., Yamamoto, R., Miyashita, K., Takahashi, Y., Ito, S., Wakamatsu, K., Mukuda, T., and Matsuura, M., 2014. Mineralized rods and cones suggest colour vision in a 300 myr-old fossil fish. *Nature Communications*, 5(1), pp. 1–6.

15. Lindgren, J., Nilsson, D. E., Sjövall, P., Jarenmark, M., Ito, S., Wakamatsu, K., Kear, B. P., Schultz, B. P., Sylvestersen, R. L., Madsen, H., and LaFountain, J. R., 2019. Fossil insect eyes shed light on trilobite optics and the arthropod pigment screen. *Nature*, 573(7772), pp. 122–125.

16. Schoenemann, B., and Clarkson, E. N., 2021. Points of view in understanding trilobite eyes. *Nature Communications*, 12(1), pp. 1–5.

17. Lindgren, J., Kuriyama, T., Madsen, H., Sjövall, P., Zheng, W., Uvdal, P., Engdahl, A., Moyer, A. E., Gren, J. A., Kamezaki, N., and Ueno, S., 2017. Biochemistry and adaptive colouration of an exceptionally preserved juvenile fossil sea turtle. *Scientific Reports*, 7(1), pp. 1–13.

18. Bustard, H. R., 1970. The adaptive significance of coloration in hatchling green sea turtles. *Herpetologica*, 26(2), p. 224–227.

19. Clements, T., Dolocan, A., Martin, P., Purnell, M. A., Vinther, J., and Gabbott, S. E., 2016. The eyes of Tullimonstrum reveal a vertebrate affinity. *Nature*, 532(7600), pp. 500–503.

20. Sallan, L., Giles, S., Sansom, R. S., Clarke, J. T., Johanson, Z., Sansom, I. J., and Janvier, P., 2017. The "Tully Monster" is not a vertebrate: Characters, convergence and taphonomy in Palaeozoic problematic animals. *Palaeontology*, 60(2), pp. 149–157.

21. McCoy, V. E., Wiemann, J., Lamsdell, J. C., Whalen, C. D., Lidgard, S., Mayer, P., Petermann, H., and Briggs, D. E., 2020. Chemical signatures of soft tissues distinguish between vertebrates and invertebrates from the Carboniferous Mazon Creek Lagerstätte of Illinois. *Geobiology*, 18, pp. 560–565.

22. Gabbott, S. E., Donoghue, P. C., Sansom, R. S., Vinther, J., Dolocan, A., and Purnell, M. A., 2016. Pigmented anatomy in Carboniferous cyclostomes and the evolution of the vertebrate eye. *Proceedings of the Royal Society B: Biological Sciences*, 283(1836), p. 2016.1151.

Chapter 5. Dino Feathers

1. Xu, X., Zhou, Z., Sullivan, C., Wang, Y., and Ren, D., 2016. An updated review of the Middle-Late Jurassic Yanliao Biota: Chronology, taphonomy, paleontology and paleoecology. *Acta Geologica Sinica-English Edition*, 90(6), pp. 2229–2243; Zhou, Z., 2014. The Jehol Biota, an Early Cretaceous terrestrial Lagerstätte: New discoveries and implications. *National Science Review*, 1(4), pp. 543–559.

2. Vinther, J., Briggs, D. E., Prum, R. O., and Saranathan, V., 2008. The colour of fossil feathers. *Biology Letters*, 4(5), pp. 522–525.

3. Carney, R. M., Vinther, J., Shawkey, M. D., D'Alba, L., and Ackermann, J., 2012. New evidence on the colour and nature of the isolated Archaeopteryx feather. *Nature Communications*, 3(1), pp. 1–6.

4. Galván, I., Jorge, A., Solano, F., and Wakamatsu, K., 2013. Vibrational characterization of pheomelanin and trichochrome F by Raman spectroscopy. *Spectrochimica Acta Part A: Molecular and Biomolecular Spectroscopy*, 110, pp. 55–59.

5. Galván, I., and Wakamatsu, K., 2016. Color measurement of the animal integument predicts the content of specific melanin forms. *RSC Advances*, 6(82), pp. 79135–79142.

6. Babarović, F., Puttick, M. N., Zaher, M., Learmonth, E., Gallimore, E. J., Smithwick, F. M., Mayr, G., and Vinther, J., 2019. Characterization of melanosomes involved in the production of non-iridescent structural feather colours and their detection in the fossil record. *Journal of the Royal Society Interface*, 16(155), p. 20180921.

7. Nordén, K. K., Faber, J. W., Babarović, F., Stubbs, T. L., Selly, T., Schiffbauer, J. D., Peharec Štefanić, P., Mayr, G., Smithwick, F. M., and Vinther, J., 2019. Melanosome diversity and convergence in the evolution of iridescent avian feathers—Implications for paleocolor reconstruction. *Evolution*, 73(1), pp. 15–27.

8. Carney, R. M., Vinther, J., Shawkey, M. D., D'Alba, L., and Ackermann, J., 2012. New evidence on the colour and nature of the isolated Archaeopteryx feather. *Nature Communications*, 3(1), pp. 1–6.

9. Li, Q., Clarke, J. A., Gao, K. Q., Zhou, C. F., Meng, Q., Li, D., D'Alba, L., and Shawkey, M. D., 2014. Melanosome evolution indicates a key physiological shift within feathered dinosaurs. *Nature*, 507(7492), pp. 350–353.

10. Rogalla, S., D'Alba, L., Verdoodt, A., and Shawkey, M. D., 2019. Hot wings: Thermal impacts of wing coloration on surface temperature during bird flight. *Journal of the Royal Society Interface*, 16(156), p. 20190032.

11. Li, Q., Gao, K. Q., Vinther, J., Shawkey, M. D., Clarke, J. A., D'Alba, L., Meng, Q., Briggs, D. E., and Prum, R. O., 2010. Plumage color patterns of an extinct dinosaur. *Science*, 327(5971), pp. 1369–1372.

12. Nordén, K. K., Faber, J. W., Babarović, F., Stubbs, T. L., Selly, T., Schiffbauer, J. D., Peharec Štefanić, P., Mayr, G., Smithwick, F. M., and Vinther, J., 2019. Melanosome diversity and convergence in the evolution of iridescent avian feathers—Implications for paleocolor reconstruction. *Evolution*, 73(1), pp. 15–27.

13. Brown, C. M., Henderson, D. M., Vinther, J., Fletcher, I., Sistiaga, A., Herrera, J., and Summons, R. E., 2017. An exceptionally preserved three-dimensional armored dinosaur reveals insights into coloration and Cretaceous predator-prey dynamics. *Current Biology*, 27(16), pp. 2514–2521.

14. Horner, J. R., and Makela, R., 1979. Nest of juveniles provides evidence of family structure among dinosaurs. *Nature, 282*(5736), pp. 296–298.

15. Wiemann, J., Yang, T. R., and Norell, M. A., 2018. Dinosaur egg colour had a single evolutionary origin. *Nature, 563*(7732), pp. 555–558.

16. Tanaka, G., Zhou, B., Zhang, Y., Siveter, D. J., and Parker, A. R., 2017. Rods and cones in an enantiornithine bird eye from the Early Cretaceous Jehol Biota. *Heliyon, 3*(12), p. e00479.

17. Schweitzer, M. H., Zheng, W., Moyer, A. E., Sjövall, P., and Lindgren, J., 2018. Preservation potential of keratin in deep time. *PLoS One, 13*(11), p. e0206569.

18. Wogelius, R. A., Manning, P. L., Barden, H. E., Edwards, N. P., Webb, S. M., Sellers, W. I., Taylor, K. G., Larson, P. L., Dodson, P., You, H., and Da-Qing, L., 2011. Trace metals as biomarkers for eumelanin pigment in the fossil record. *Science, 333*(6049), pp. 1622–1626.

Chapter 6. Ancient Biometals

1. Greenwalt, D., Rose, T. R., and Chatzimanolis, S., 2016. Preservation of mandibular zinc in a beetle from the Eocene Kishenehn Formation of Montana, USA. *Canadian Journal of Earth Sciences, 53*(6), pp. 614–621.

2. Chou, C. C., Martin-Martinez, F. J., Qin, Z., Dennis, P. B., Gupta, M. K., Naik, R. R., and Buehler, M. J., 2017. Ion effect and metal-coordinated cross-linking for multiscale design of nereis jaw inspired mechanomutable materials. *ACS Nano, 11*(2), pp. 1858–1868.

3. Chen, J., Ohta, Y., Nakamura, H., Masunaga, H., and Numata, K., 2021. Aqueous spinning system with a citrate buffer for highly extensible silk fibers. *Polymer Journal, 53*(1), pp. 179–189.

4. Edwards, N. P., Wogelius, R. A., Bergmann, U., Larson, P., Sellers, W. I., and Manning, P. L., 2013. Mapping prehistoric ghosts in the synchrotron. *Applied Physics A, 111*(1), pp. 147–155.

5. Rossi, V., McNamara, M. E., Webb, S. M., Ito, S., and Wakamatsu, K., 2019. Tissue-specific geometry and chemistry of modern and fossilized melanosomes reveal internal anatomy of extinct vertebrates. *Proceedings of the National Academy of Sciences, 116*(36), pp. 17880–17889.

6. Terrill, D. F., Henderson, C. M., and Anderson, J. S., 2018. New applications of spectroscopy techniques reveal phylogenetically significant soft tissue residue in Paleozoic conodonts. *Journal of Analytical Atomic Spectrometry, 33*(6), pp. 992–1002.

7. Briggs, D. E., Clarkson, E. N., and Aldridge, R. J., 1983. The conodont animal. *Lethaia, 16*(1), pp. 1–14.

8. Wogelius, R. A., Manning, P. L., Barden, H. E., Edwards, N. P., Webb, S. M., Sellers, W. I., Taylor, K. G., Larson, P. L., Dodson, P., You, H., and Da-Qing, L., 2011. Trace metals as biomarkers for eumelanin pigment in the fossil record. *Science*, *333*(6049), pp. 1622–1626.

9. Edwards, N. P., Manning, P. L., Bergmann, U., Larson, P. L., Van Dongen, B. E., Sellers, W. I., Webb, S. M., Sokaras, D., Alonso-Mori, R., Ignatyev, K., Barden, H. E., van Veelen, A., Anne, J., Egerton, V. M., and Wogelius, R.A., 2014. Leaf metallome preserved over 50 million years. *Metallomics*, *6*(4), pp. 774–782.

Chapter 7. Proteins and Proteomes

1. Von Koenigswald, G.H.R., 1935. Eine fossile Saugetierfauna mit Simia aus Sudchina. In *Proc. Kon. Ned. Akad. Wetensch.*, *38*, pp. 872–879.

2. Zuckerkandl, E., and Pauling, L., 1965. "Evolutionary Divergence and Convergence in Proteins," in Vernon Bryson and Henry Vogel, eds., *Evolving Genes and Proteins*. Academic Press, pp. 97–166.

3. http://scarc.library.oregonstate.edu/coll/pauling/blood/notes/sci6.007.17-notes-waddington.html.

4. Stedman, H. H., Kozyak, B. W., Nelson, A., Thesier, D. M., Su, L. T., Low, D. W., Bridges, C. R., Shrager, J. B., Minugh-Purvis, N., and Mitchell, M. A., 2004. Myosin gene mutation correlates with anatomical changes in the human lineage. *Nature*, *428*(6981), pp. 415–418.

5. Wang, W., 2009. New discoveries of *Gigantopithecus blacki* teeth from Chuifeng Cave in the Bubing Basin, Guangxi, south China. *Journal of Human Evolution*, *57*(3), pp. 229–240.

6. Welker, F., Ramos-Madrigal, J., Kuhlwilm, M., Liao, W., Gutenbrunner, P., de Manuel, M., Samodova, D., Mackie, M., Allentoft, M. E., Bacon, A. M., and Collins, M. J., 2019. Enamel proteome shows that Gigantopithecus was an early diverging pongine. *Nature*, *576*(7786), pp. 262–265.

7. Buckley, M., 2013. A molecular phylogeny of Plesiorycteropus reassigns the extinct mammalian order "Bibymalagasia." *PLoS One*, *8*(3), p. e59614.

8. Jiang, X., Ye, M., Jiang, X., Liu, G., Feng, S., Cui, L., and Zou, H., 2007. Method development of efficient protein extraction in bone tissue for proteome analysis. *Journal of Proteome Research*, *6*(6), pp. 2287–2294.

9. Cappellini, E., Welker, F., Pandolfi, L., Madrigal, J. R., Fotakis, A. K., Lyon, D., Mayar, V.J.M., Bukhsianidze, M., Jersie-Christensen, R. R., Mackie, M., and Ginolhac, A., 2018. Early Pleistocene enamel proteome sequences from Dmanisi resolve *Stephanorhinus* phylogeny. *BioRxiv*, 407692.

10. Cappellini, E., Jensen, L. J., Szklarczyk, D., Ginolhac, A., da Fonseca, R. A., Stafford Jr, T. W., Holen, S. R., Collins, M. J., Orlando, L., Willerslev, E., and Gilbert,

M.T.P., 2012. Proteomic analysis of a pleistocene mammoth femur reveals more than one hundred ancient bone proteins. *Journal of Proteome Research*, 11(2), pp. 917–926.

11. Orlando, L., Ginolhac, A., Zhang, G., Froese, D., Albrechtsen, A., Stiller, M., Schubert, M., Cappellini, E., Petersen, B., Moltke, I., Johnson, P.L.F., Fumagalli, M., Vilstrup, J. T., Raghavan, M., Korneliussen, T. S., Sapfo Malaspinas, A. S., Korbinian Vogt, J. K., Szklarczyk, D., Kelstrup, C., Vinther, J., Dolocan, A., Stenderup, J., Velazquez, A.M.V., Cahill, J., Rasmussen, M., Wang, X., Min, J., Zazula, G. D., Seguin-Orlando, A., Mortensen, C., Magnussen, K., Thompson, J. F., Weinstock, J., Gregersen, M. K., Røed, Eisenmann, V., Rubin, C. J., Miller, D. C., Antczak, D. F., Bertelsen, M., Brunak, S., Al-Rasheid, K. A. S., Ryder, O., Andersson, L., Mundy, J., Krogh, A., Gilbert, M.T.P., Kjær, K. H., Sicheritz-Pontén, T., Jensen, L. J., Olsen, J., Hofreiter, M., Nielsen, R., Shapiro, B., Wang, J., and Willerslev, E. 2013. Recalibrating Equus evolution using the genome sequence of an early Middle Pleistocene horse. *Nature*, 499(7456), pp. 74–78.

12. Hill, R. C., Wither, M. J., Nemkov, T., Barrett, A., D'Alessandro, A., Dzieciatkowska, M., and Hansen, K. C., 2005. Preserved proteins from extinct *Bison Latifrons* identified by tandem mass spectrometry; hydroxylysine glycosides are a common feature of ancient collagen. *Molecular & Cellular Proteomics*. 14(7), pp. 1946–58; Wadsworth, C., and Buckley, M., 2014. Proteome degradation in fossils: Investigating the longevity of protein survival in ancient bone. *Rapid Communications in Mass Spectrometry*, 28(6), pp. 605–615; Waters, M. R., and Stafford, T. W., 2007. Redefining the age of Clovis: 743 implications for the peopling of the Americas. *Science*, 315(5815), pp. 1122–1126.

13. Ehrlich, H., Rigby, J. K., Botting, J. P., Tsurkan, M. V., Werner, C., Schwille, P., Petrášek, Z., Pisera, A., Simon, P., Sivkov, V. N., Vyalikh, D. V., Molodtsov, S. L., Kurek, D., Kammer, M., Hunoldt, S., Born, R., Stawski, D., Steinhof, A., Bazhenov, V. V., and Geisler, T., 2013. Discovery of 505-million-year-old chitin in the basal demosponge *Vauxia gracilenta*. *Scientific Reports*, 3(1), pp. 1–6.

14. Buckley, M., Collins, M., and Thomas-Oates, J., 2008. A method of isolating the collagen (I) α2 chain carboxytelopeptide for species identification in bone fragments. *Analytical Biochemistry*, 374(2), pp. 325–334.

15. Wilson, M., and Churcher, C. S., 1978. Late Pleistocene *Camelops* from the Gallelli Pit, Calgary, Alberta: Morphology and geologic setting. *Canadian Journal of Earth Sciences*, 15(5), pp. 729–740.

16. Rybczynski, N., Gosse, J. C., Harington, C. R., Wogelius, R. A., Hidy, A. J., and Buckley, M., 2013. Mid-Pliocene warm-period deposits in the High Arctic yield insight into camel evolution. *Nature Communications*, 4(1), pp. 1–9.

17. Buckley, M., Lawless, C., and Rybczynski, N., 2019. Collagen sequence analysis of fossil camels, Camelops and cf Paracamelus, from the Arctic and sub-Arctic of Plio-Pleistocene North America. *Journal of Proteomics*, 194, pp. 218–225.

18. Buckley, M., Harvey, V. L., and Chamberlain, A. T., 2017. Species identification and decay assessment of Late Pleistocene fragmentary vertebrate remains from Pin Hole Cave (Creswell Crags, UK) using collagen fingerprinting. *Boreas, 46*(3), pp. 402–411.

19. Gillespie, J. M., 1970. Mammoth hair: Stability of α-keratin structure and constituent proteins. *Science, 170*(3962), pp. 1100–1102.

20. Strasser, B., Mlitz, V., Hermann, M., Rice, R. H., Eigenheer, R. A., Alibardi, L., Tschachler, E., and Eckhart, L., 2014. Evolutionary origin and diversification of epidermal barrier proteins in amniotes. *Molecular Biology and Evolution, 31*(12), pp. 3194–3205.

21. Drake, J. L., Whitelegge, J. P., and Jacobs, D. K., 2020. First sequencing of ancient coral skeletal proteins. *Scientific Reports, 10*(1), pp. 1–11.

22. Demarchi, B., Hall, S., Roncal-Herrero, T., Freeman, C. L., Woolley, J., Crisp, M. K., Wilson, J., Fotakis, A., Fischer, R., Kessler, B. M., and Jersie-Christensen, R. R., 2016. Protein sequences bound to mineral surfaces persist into deep time. *eLife, 5*, p. e17092.

23. Nerlich, A. G., Fleckinger, A., and Peschel, O., 2020. Life and diseases of the neolithic glacier mummy "Ötzi." In Shin, D. H., and Bianucci, R. (eds.), *The Handbook of Mummy Studies: New Frontiers in Scientific and Cultural Perspectives*, Springer, pp. 1–22.

24. Lippert, A., Gostner, P., Egarter-Vigl, E., and Pernter, P., 2007. Vom Leben und Sterben des Ötztaler Gletschermannes: Neue medizinische und archäologische Erkenntnisse. *Germania: Anzeiger der Römisch-Germanischen Kommission des Deutschen Archäologischen Instituts, 85*(1), pp. 1–21.

25. Maixner, F., Overath, T., Linke, D., Janko, M., Guerriero, G., van den Berg, B. H., Stade, B., Leidinger, P., Backes, C., Jaremek, M., Kneissl, B., Meder, B., Franke, A., Egarter-Vigl, E., Meese, E., Schwarz, A., Tholey, A., Zink, A., and Keller, A. 2013. Paleoproteomic study of the Iceman's brain tissue. *Cellular and mMolecular Life Sciences, 70*(19), pp. 3709–3722.

Chapter 8. Dino Bones

1. Service, R. F. 2017. "I don't care what they say about me": Paleontologist stares down critics in her hunt for dinosaur proteins. *Science*, Sept. 13. https://doi.org/10.1126/science.aap9404.

2. Schweitzer, M. H., Johnson, C., Zocco, T. G., Horner, J. R., and Starkey, J. R., 1997. Preservation of biomolecules in cancellous bone of Tyrannosaurus rex. *Journal of Vertebrate Paleontology, 17*(2), pp. 349–359.

3. Schweitzer, M. H., Marshall, M., Carron, K., Bohle, D. S., Busse, S. C., Arnold, E. V., Barnard, D., Horner, J. R., and Starkey, J. R., 1997. Heme compounds in dinosaur trabecular bone. *Proceedings of the National Academy of Sciences, 94*(12), pp. 6291–6296.

4. Schweitzer, M. H., Watt, J. A., Avci, R., Knapp, L., Chiappe, L., Norell, M., and Marshall, M., 1999. Beta-keratin specific immunological reactivity in feather-like structures of the Cretaceous Alvarezsaurid, *Shuvuuia deserti. Journal of Experimental Zoology, 285*(2), pp. 146–157; Schweitzer, M. H., Watt, J. A., Avci, R., Forster, C. A., Krause, D. W., Knapp, L., Rogers, R. R., Beech, I., and Marshall, M., 1999. Keratin immunoreactivity in the Late Cretaceous bird *Rahonavis ostromi. Journal of Vertebrate Paleontology, 19*(4), pp. 712–722.

5. Schweitzer, M. H., Wittmeyer, J. L., Horner, J. R., and Toporski, J. K., 2005. Soft-tissue vessels and cellular preservation in *Tyrannosaurus rex. Science, 307*(5717), pp. 1952–1955.

6. Wiemann, J., Fabbri, M., Yang, T. R., Stein, K., Sander, P. M., Norell, M. A., and Briggs, D. E., 2018. Fossilization transforms vertebrate hard tissue proteins into N-heterocyclic polymers. *Nature Communications, 9*(1), pp. 1–9.

7. Pawlicki, R., and Nowogrodzka-Zagórska, M., 1998. Blood vessels and red blood cells preserved in dinosaur bones. *Annals of Anatomy-Anatomischer Anzeiger, 180*(1), pp. 73–77.

8. Pawlicki, R., and Nowogrodzka-Zagórska, M., 1998. Blood vessels and red blood cells preserved in dinosaur bones. *Annals of Anatomy-Anatomischer Anzeiger, 180*(1), pp. 73–77.

9. Asara, J. M., Schweitzer, M. H., Phillips, M. P., Freimark, L. M., and Cantley, L. C. 2007. Protein sequences from mastodon (*Mammut americanum*) and dinosaur (*Tyrannosaurus rex*) revealed by mass spectrometry. *Science, 316*, pp. 280–285.

10. Schweitzer, M. H., Schroeter, E. R., and Goshe, M. B., 2014. Protein molecular data from ancient (> 1 million years old) fossil material: Pitfalls, possibilities and grand challenges. *Analytical Chemistry, 86*(14), pp. 6731–6740.

11. Schroeter, E. R., DeHart, C. J., Cleland, T. P., Zheng, W., Thomas, P. M., Kelleher, N. L., Bern, M., and Schweitzer, M. H., 2017. Expansion for the *Brachylophosaurus canadensis* collagen I sequence and additional evidence of the preservation of Cretaceous protein. *Journal of Proteome Research, 16*(2), pp. 920–932.

12. Lindgren, J., Uvdal, P., Engdahl, A., Lee, A. H., Alwmark, C., Bergquist, K. E., Nilsson, E., Ekström, P., Rasmussen, M., Douglas, D. A., and Polcyn, M. J., 2011. Microspectroscopic evidence of Cretaceous bone proteins. *PLoS One, 6*(4), p. e19445; Lindgren, J., Sjövall, P., Thiel, V., Zheng, W., Ito, S., Wakamatsu, K., Hauff, R., Kear, B. P., Engdahl, A., Alwmark, C., and Eriksson, M. E., 2018. Soft-tissue evidence for homeothermy and crypsis in a Jurassic ichthyosaur. *Nature, 564*(7736), pp. 359–365.

13. Embery, G., Milner, A. C., Waddington, R .J., Hall, R. C., Langley, M. S., and Milan, A. M., 2003. Identification of proteinaceous material in the bone of the dinosaur Iguanodon. *Connective Tissue Research, 44*(1), pp. 41–46.

14. Service, R. F. 2017. "I don't care what they say about me": Paleontologist stares down critics in her hunt for dinosaur proteins. *Science*, Sept. 13. https://doi.org/10.1126/science.aap9404.

15. Kaye, T. G., Gaugler, G., and Sawlowicz, Z., 2008. Dinosaurian soft tissues interpreted as bacterial biofilms. *PLoS One*, 3(7), p. e2808.

16. Lindgren, J., Uvdal, P., Engdahl, A., Lee, A. H., Alwmark, C., Bergquist, K. E., Nilsson, E., Ekström, P., Rasmussen, M., Douglas, D. A., and Polcyn, M. J., 2011. Microspectroscopic evidence of Cretaceous bone proteins. *PLoS One*, 6(4), p. e19445.

17. Schweitzer, M. H., Moyer, A. E., and Zheng, W., 2016. Testing the hypothesis of biofilm as a source for soft tissue and cell-like structures preserved in dinosaur bone. *PLoS One*, 11(2), p. e0150238.

18. Saitta, E. T., Liang, R., Lau, M. C., Brown, C. M., Longrich, N. R., Kaye, T. G., Novak, B. J., Salzberg, S. L., Norell, M. A., Abbott, G. D., and Dickinson, M. R., 2019. Cretaceous dinosaur bone contains recent organic material and provides an environment conducive to microbial communities. *eLife*, 8, p. e46205.

19. Armitage, M. H., 2016. Preservation of *Triceratops horridus* tissue cells from the Hell Creek Formation, MT. *Microscopy Today*, 24(1), pp. 18–23.

20. Armitage, M. H. Bacteria trapped in a Nanotyrannus vein valve come back to life, Part 2. https://www.youtube.com/watch?v=DFK2kqR41mI. Unpublished.

21. Hodge, J. E., 1953. Dehydrated foods, chemistry of browning reactions in model systems. *Journal of Agricultural and Food Chemistry*, 1(15), pp. 928–943.

22. Tareke, E., Rydberg, P., Karlsson, P., Eriksson, S., and Törnqvist, M., 2002. Analysis of acrylamide, a carcinogen formed in heated foodstuffs. *Journal of Agricultural and Food Chemistry*, 50(17), pp. 4998–5006.

23. Wiemann, J., Fabbri, M., Yang, T. R., Stein, K., Sander, P. M., Norell, M. A., and Briggs, D. E., 2018. Fossilization transforms vertebrate hard tissue proteins into N-heterocyclic polymers. *Nature Communications*, 9(1), pp. 1–9.

24. Norell, M. A., Wiemann, J., Fabbri, M., Yu, C., Marsicano, C. A., Moore-Nall, A., Varricchio, D. J., Pol, D., and Zelenitsky, D. K., 2020. The first dinosaur egg was soft. *Nature*, 583(7816), pp. 406–410.

25. Saitta, E. T., Rogers, C., Brooker, R. A., Abbott, G. D., Kumar, S., O'Reilly, S. S., Donohoe, P., Dutta, S., Summons, R. E., and Vinther, J., 2017. Low fossilization potential of keratin protein revealed by experimental taphonomy. *Palaeontology*, 60(4), pp. 547–556; Saitta, E. T., and Vinther, J., 2019. A perspective on the evidence for keratin protein preservation in fossils: An issue of replication versus validation. *Palaeontologia Electronica*, 22.

26. Greenwold, M. J., and Sawyer, R. H., 2011. Linking the molecular evolution of avian beta (β) keratins to the evolution of feathers. *Journal of Experimental Zoology Part B: Molecular and Developmental Evolution*, 316(8), pp. 609–616.

27. Pan, Y., et al., 2019. The molecular evolution of feathers with direct evidence from fossils. *Proceedings of the National Academy of Sciences*, 116(8), pp. 3018–3023.

28. Alleon, J., Montagnac, G., Reynard, B., Brulé, T., Thoury, M., and Gueriau, P., 2021. Pushing Raman spectroscopy over the edge: Purported signatures of organic molecules in fossil animals are instrumental artefacts. *BioEssays*, 43(4), p. 2000295.

29. Bretz, J. H., 1923. The channeled scablands of the Columbia Plateau. *Journal of Geology*, 31(8), pp. 617–649.

30. Bailleul, A. M., Zheng, W., Horner, J. R., Hall, B. K., Holliday, C. M., and Schweitzer, M. H., 2020. Evidence of proteins, chromosomes and chemical markers of DNA in exceptionally preserved dinosaur cartilage. *National Science Review*, 7(4), pp. 815–822.

31. Woodward, S. R., Weyand, N. J., and Bunnell, M., 1994. DNA sequence from Cretaceous period bone fragments. *Science*, 266(5188), pp. 1229–1232.

32. Hedges, S. B., and Schweitzer, M. H., 1995. Detecting dinosaur DNA. *Science*, 268, p. 1191.

Chapter 9. Ancient DNA's Tenuous Origins

1. Poinar, G. O., and Hess, R., 1982. Ultrastructure of 40-million-year-old insect tissue. *Science*, 215(4537), pp. 1241–1242.

2. Cano, R. J., Poinar, H. N., Pieniazek, N. J., Acra, A., and Poinar, G. O., 1993. Amplification and sequencing of DNA from a 120–135-million-year-old weevil. *Nature*, 363(6429), pp. 536–538; Cano, R. J., Poinar, H., and Poinar Jr, G. O., 1992. Isolation and partial characterization of DNA from the bee Proplebeia dominicana (Apidae: Hymenoptera) in 25–40-million-year-old amber. *Med. Sci. Res*, 20, pp. 249–251; DeSalle, R., Gatesy, J., Wheeler, W., and Grimaldi, D., 1992. DNA sequences from a fossil termite in Oligo-Miocene amber and their phylogenetic implications. *Science*, 257(5078), pp. 1933–1936; Weyand, N. J., and Bunnell, M., 1994. DNA sequence from Cretaceous period bone fragments. *Science*, 266(5188), pp. 1229–1232.

3. Austin, J. J., Smith, A. B., Fortey, R. A., and Thomas, R. H., 1998. Ancient DNA from amber inclusions: A review of the evidence. *Ancient Biomolecules*, 2(2), pp. 167–176.

4. Penney, D., Wadsworth, C., Fox, G., Kennedy, S. L., Preziosi, R. F., and Brown, T. A., 2013. Absence of ancient DNA in sub-fossil insect inclusions preserved in "Anthropocene" Colombian copal. *PLoS One*, 8(9), p. e73150.

5. Thomsen, P. F., Elias, S., Gilbert, M.T.P., Haile, J., Munch, K., Kuzmina, S., Froese, D. G., Sher, A., Holdaway, R. N., and Willerslev, E., 2009. Non-destructive sampling of ancient insect DNA. *PLoS One*, 4(4), p. e5048.

6. Stankiewicz, B. A., Poinar, H. N., Briggs, D. E., Evershed, R. P., and Poinar Jr, G. O., 1998. Chemical preservation of plants and insects in natural resins. *Proceedings of the Royal Society of London. Series B: Biological Sciences*, 265(1397), pp. 641–647.

7. Kowalewska, M., and Szwedo, J., 2009. Examination of the Baltic amber inclusion surface using SEM techniques and X-ray microanalysis. *Palaeogeography, Palaeoclimatology, Palaeoecology, 271*(3–4), pp. 287–291.

8. Liepelt, S., Sperisen, C., Deguilloux, M. F., Petit, R. J., Kissling, R., Spencer, M., De Beaulieu, J. L., Taberlet, P., Gielly, L., and Ziegenhagen, B., 2006. Authenticated DNA from ancient wood remains. *Annals of Botany, 98*(5), pp. 1107–1111.

9. Cooper, A., and Poinar, H. N., 2000. Ancient DNA: Do it right or not at all. *Science, 289*(5482), pp. 1139–1139.

10. Mullis, K. B., 1990. The unusual origin of the polymerase chain reaction. *Scientific American, 262*(4), pp. 56–65.

11. Brock, T. D., and H. Freeze, 1969. *Thermus aquaticus* gen. n. and sp. N., a non-sporulating extreme thermophile. *J. Bacteriology, 98*, pp. 289–297.

12. Brock, T. D., 1995. The road to Yellowstone—and beyond. *Annual Review Microbiology, 49*, pp. 1–28.

13. Pääbo, S., 1985. Molecular cloning of ancient Egyptian mummy DNA. *Nature, 314*(6012), pp. 644–645. For a more recent reference, see Schuenemann, V. J., Peltzer, A., Welte, B., Van Pelt, W. P., Molak, M., Wang, C. C., Furtwängler, A., Urban, C., Reiter, E., Nieselt, K., Teßmann, B., Francken, M., Harvati, K., Haak, W., Schiffels, S., and Krause, J., 2017. Ancient Egyptian mummy genomes suggest an increase of Sub-Saharan African ancestry in post-Roman periods. *Nature Communications, 8*(1), pp. 1–11; Higuchi, R., Bowman, B., Freiberger, M., Ryder, O. A., and Wilson, A. C., 1984. DNA sequences from the quagga, an extinct member of the horse family. *Nature, 312*(5991), pp. 282–284.

14. Lister, A. M., and Sher, A. V., 2015. Evolution and dispersal of mammoths across the Northern Hemisphere. *Science, 350*(6262), pp. 805–809.

15. Van der Valk, T., Pečnerová, P., Díez-del-Molino, D., Bergström, A., Oppenheimer, J., Hartmann, S., Xenikoudakis, G., Thomas, J. A., Dehasque, M., Sağlıcan, E., and Fidan, F. R., 2021. Million-year-old DNA sheds light on the genomic history of mammoths. *Nature, 591*(7849), pp. 265–269.

16. Orlando, L., Ginolhac, A., Zhang, G., Froese, D., Albrechtsen, A., Stiller, M., Schubert, M., Cappellini, E., Petersen, B., Moltke, I., and Johnson, P. L., 2013. Recalibrating Equus evolution using the genome sequence of an early Middle Pleistocene horse. *Nature, 499*(7456), pp. 74–78.

17. Pečnerová, P., Díez-del-Molino, D., Vartanyan, S., and Dalén, L., 2016. Changes in variation at the MHC class II DQA locus during the final demise of the woolly mammoth. *Scientific Reports, 6*(1), pp. 1–11.

18. Barlow, A, Paijmans, J. L., Alberti, F., Gasparyan, B., Bar-Oz, G., Pinhasi, R., Foronova, I., Puzachenko, A. Y., Pacher, M., Dalén, L., and Baryshnikov, G., 2021. Middle Pleistocene genome calibrates a revised evolutionary history of extinct cave bears. *Current Biology, 31*(8), pp. 1771–1779.e7.

19. Lederberg, J., 1952. Cell genetics and hereditary symbiosis. *Physiological Reviews*, 32(4), pp. 403–430.

20. Campbell, K. L., Roberts, J. E., Watson, L. N., Stetefeld, J., Sloan, A. M., Signore, A. V., Howatt, J. W., Tame, J. R., Rohland, N., Shen, T. J., and Austin, J. J., 2010. Substitutions in woolly mammoth hemoglobin confer biochemical properties adaptive for cold tolerance. *Nature Genetics*, 42(6), pp. 536–540.

21. Vilstrup, J. T., Seguin-Orlando, A., Stiller, M., Ginolhac, A., Raghavan, M., Nielsen, S. C., Weinstock, J., Froese, D., Vasiliev, S. K., Ovodov, N. D., and Clary, J., 2013. Mitochondrial phylogenomics of modern and ancient equids. *PLoS One*, 8(2), p. e55950.

22. Rasmussen, S., Allentoft, M., Nielsen, K., Orlando, L., Sikora, M., Pedersen, A., Schubert, M., VanáDam, A., Kapel, C., Avetisyan, P., and Epimakhov, A., 2015. Early divergent strains of *Yersinia pestis* in Eurasia 5,000 years ago. *Cell*, 163(3), pp. 571–582.

23. Zimbler, D. L., Schroeder, J. A., Eddy, J. L., and Lathem, W. W., 2015. Early emergence of *Yersinia pestis* as a severe respiratory pathogen. *Nature Communications*, 6, p. 7487.

24. Golenberg, E. M., Giannasi, D. E., Clegg, M. T., Smiley, C. J., Durbin, M., Henderson, D., and Zurawski, G., 1990. Chloroplast DNA sequence from a Miocene *Magnolia* species. *Nature*, 344(6267), pp. 656–658.

25. Kawamuro, K., Kinoshita, I., Suyama, Y., and Takahara, H., 1995. Inspection of DNA in fossil pollen of *Abies* spp. From late Pleistocene peat. *Journal of the Japanese Forestry Society*, 77(3), pp. 272–274.

26. Liepelt, S., Sperisen, C., Deguilloux, M. F., Petit, R. J., Kissling, R., Spencer, M., De Beaulieu, J. L., Taberlet, P., Gielly, L., and Ziegenhagen, B., 2006. Authenticated DNA from ancient wood remains. *Annals of Botany*, 98(5), pp. 1107–1111.

27. Murray, D. C., Pearson, S. G., Fullagar, R., Chase, B. M., Houston, J., Atchison, J., White, N. E., Bellgard, M. I., Clarke, E., Macphail, M., and Gilbert, M.T.P., 2012. High-throughput sequencing of ancient plant and mammal DNA preserved in herbivore middens. *Quaternary Science Reviews*, 58, pp. 135–145.

28. Gravendeel, B., Protopopov, A., Bull, I., Duijm, E., Gill, F., Nieman, A., Rudaya, N., Tikhonov, A. N., Trofimova, S., van Reenen, G. B., and Vos, R., 2014. Multiproxy study of the last meal of a mid-Holocene Oyogos Yar horse, Sakha Republic, Russia. *The Holocene*, 24(10), pp. 1288–1296; Willerslev, E., Davison, J., Moora, M., Zobel, M., Coissac, E., Edwards, M. E., Lorenzen, E. D., Vestergård, M., Gussarova, G., Haile, J., Craine, J., Bergmann, G., Gielly, L., Boessenkool, S., Epp, L. S., Pearman, P. B., Cheddadi, R., Murray, D., Bråthen, K. A., Yoccoz, N., Binney, H., Cruaud, C., Wincker, P., Goslar, T., Alsos, I. G., Bellemain, E., Brysting, A. K., Elven, R., Sønstebø, J. H., Murton, J., Sher, A., Rasmussen, M., Rønn, R., Mourier, T., Cooper, A., Austin, J., Möller, P., Froese, D., Zazula, G., Pompanon, F., Rioux, D.,

Niderkorn, V., Tikhonov, A., Savvinov, G., Roberts, R. G., MacPhee, R.D.E., Gilbert, M.T.P., Kjær, K. H., Orlando,L., Brochmann, C., and Taberle, P., 2014. Fifty thousand years of Arctic vegetation and megafaunal diet. *Nature, 506*(7486), pp. 47–51; Hofreiter, M., Poinar, H. N., Spaulding, W. G., Bauer, K., Martin, P. S., Possnert, G., and Pääbo, S., 2000. A molecular analysis of ground sloth diet through the last glaciation. *Molecular Ecology, 9*(12), pp. 1975–1984; van Geel, B., Aptroot, A., Baittinger, C., Birks, H. H., Bull, I. D., Cross, H. B., Evershed, R. P., Gravendeel, B., Kompanje, E. J., Kuperus, P., and Mol, D., 2008. The ecological implications of a Yakutian mammoth's last meal. *Quaternary Research, 69*(3), pp. 361–376.

29. Rollo, F., Ubaldi, M., Ermini, L., and Marota, I., 2002. Ötzi's last meals: DNA analysis of the intestinal content of the Neolithic glacier mummy from the Alps. *Proceedings of the National Academy of Sciences, 99*(20), pp. 12594–12599.

30. Soulié-Märsche, I., and García, A., 2015. Gyrogonites and oospores, complementary viewpoints to improve the study of the charophytes (Charales). *Aquatic Botany, 120*, pp. 7–17.

31. Willerslev, E., Cappellini, E., Boomsma, W., Nielsen, R., Hebsgaard, M. B., Brand, T. B., Hofreiter, M., Bunce, M., Poinar, H. N., Dahl-Jensen, D., Johnsen, S., Steffensen, J. P., Bennike, O., Schwenninger, J-L., Nathan, R., Armitage, S., Hoog, C-J. D., Alfimov, V., Christl, M., Beer, J., Muscheler, R., Barker, J., Sharp, M., Penkman, K.E.H., Haile, J., Taberlet, P., Gilbert, M.T.P., Casoli, A., Campani, E., and Collins, M. J. 2007. Ancient biomolecules from deep ice cores reveal a forested southern Greenland. *Science, 317*(5834), pp. 111–114.

32. LeBlanc, S. A., Cobb Kreisman, L. S., Kemp, B. M., Smiley, F. E., Carlyle, S. W., Dhody, A. N., and Benjamin, T., 2007. Quids and aprons: Ancient DNA from artifacts from the American Southwest. *Journal of Field Archaeology, 32*(2), pp. 161–175.

33. Morard, R., Lejzerowicz, F., Darling, K. F., Lecroq-Bennet, B., Winther Pedersen, M., Orlando, L., Pawlowski, J., Mulitza, S., Vargas, C. D., and Kucera, M., 2017. Planktonic foraminifera-derived environmental DNA extracted from abyssal sediments preserves patterns of plankton macroecology. *Biogeosciences, 14*(11), pp. 2741–2754.

34. Jensen, T. Z., Niemann, J., Iversen, K. H., Fotakis, A. K., Gopalakrishnan, S., Vågene, Å. J.,Pedersen, M. W., Sinding, M.H.S., Ellegaard, M. R., Allentoft, M. E., Lanigan, L. T., Taurozzi, A. J., Holtsmark Nielsen, S. H., Dee, M. W., Mortensen, M. N., Christensen, M. C., Sørensen, S. A., Collins, M. J., Gilbert, M.T.P., Sikora, M., Rasmussen, S., and Schroeder, H. 2019. A 5,700-year-old human genome and oral microbiome from chewed birch pitch. *Nature Communications, 10*(1), pp. 1–10.

35. Kashuba, N., Kirdök, E., Damlien, H., Manninen, M. A., Nordqvist, B., Persson, P., and Götherstörm, A., 2018. Ancient DNA from chewing gums connects material culture and genetics of Mesolithic hunter-gatherers in Scandinavia. *bioRxiv*, p. 485045.

36. Slon, V., Hopfe, C., Weiß, C. L., Mafessoni, F., De La Rasilla, M., Lalueza-Fox, C., Rosas, A., Soressi, M., Knul, M. V., Miller, R., Stewart, J. R., Derevianko, A. P., Jacobs, Z., Li, B., Roberts, R. G., Shunkov, M. V., Lumley, H. D., Perrenoud, C., Gušić, I., Kućan, Z., Rudan, P., Aximu-Petri, A., Essel, E., Nagel, S., Nickel, B., Schmidt, A., Prüfer, K., Kelso, J., Burbano, H. A., Pääbo, S., and Meyer, M. 2017. Neandertal and Denisovan DNA from Pleistocene sediments. *Science*, *356*(6338), pp. 605–608.

Chapter 10. Our Inner Neandertal

1. Pääbo, S., 2014. *Neanderthal Man: In Search of Lost Genomes*. Hachette UK.

2. Churchill, S. E., 2006. Bioenergetic perspectives on Neanderthal thermoregulatory and activity budgets. In *Neanderthals Revisited: New Approaches and Perspectives*. Springer, Dordrecht, pp. 113–133.

3. Retshabile, G., Mlotshwa, B. C., Williams, L., Mwesigwa, S., Mboowa, G., Huang, Z., Rustagi, N., Swaminathan, S., Katagirya, E., Kyobe, S., and Wayengera, M., 2018. Whole-exome sequencing reveals uncaptured variation and distinct ancestry in the southern African population of Botswana. *American Journal of Human Genetics*, *102*(5), pp. 731–743; Choudhury, A., Aron, S., Botigué, L. R., Sengupta, D., Botha, G., Bensellak, T., Wells, G., Kumuthini, J., Shriner, D., Fakim, Y. J., and Ghoorah, A. W., 2020. High-depth African genomes inform human migration and health. *Nature*, *586*(7831), pp. 741–748.

4. Ferring, R., Oms, O., Agustí, J., Berna, F., Nioradze, M., Shelia, T., Tappen, M., Vekua, A., Zhvania, D., and Lordkipanidze, D., 2011. Earliest human occupations at Dmanisi (Georgian Caucasus) dated to 1.85–1.78 Ma. *Proceedings of the National Academy of Sciences*, *108*(26), pp. 10432–10436.

5. Rizal, Y., Westaway, K. E., Zaim, Y., van den Bergh, G. D., Bettis, E. A., Morwood, M. J., Huffman, O. F., Grün, R., Joannes-Boyau, R., Bailey, R. M., and Westaway, M. C., 2020. Last appearance of *Homo erectus* at Ngandong, Java, 117,000–108,000 years ago. *Nature*, *577*(7790), pp. 381–385.

6. Li, H., Mallick, S., and Reich, D., 2014. Population size changes and split times. Supplementary Information 12. In Prüfer, K., Racimo, F., Patterson, N., Jay, F., Sankararaman, S., Sawyer, S., Heinze, A., Renaud, G., Sudmant, P. H., de Filippo, C., Li, H., Mallick, S., Dannemann, M., Fu, Q., Kircher, M., Kuhlwilm, M., Lachmann, M., Meyer, M., Ongyerth, M., Siebauer, M., Theunert, C., Tandon, A., Moorjani, P., Pickrell, J., Mullikin, J. C., Vohr, S. H., Green, R. E., Hellmann, I., Johnson, P.L.F., Blanche, H., Cann, H., Kitzman, J. O., Shendure, J., Eichler, E. E., Lein, E. S., Bakken, T. E., Golovanova, L. V., Doronichev, V. B., Shunkov, M. V., Derevianko, A. P., Viola, B., Slatkin, M., Reich, D., Kelso, J., and Pääbo., S., 2014. The complete genome sequence of a Neanderthal from the Altai Mountains. *Nature*, *505*, pp. 43–49.

7. Hublin, J. J., Ben-Ncer, A., Bailey, S. E., Freidline, S. E., Neubauer, S., Skinner, M. M., Bergmann, I., Le Cabec, A., Benazzi, S., Harvati, K., and Gunz, P., 2017. New fossils from Jebel Irhoud, Morocco and the pan-African origin of *Homo sapiens*. *Nature*, 546(7657), pp. 289–292; Richter, D., Grün, R., Joannes-Boyau, R., Steele, T. E., Amani, F., Rué, M., Fernandes, P., Raynal, J. P., Geraads, D., Ben-Ncer, A., and Hublin, J. J., 2017. The age of the hominin fossils from Jebel Irhoud, Morocco, and the origins of the Middle Stone Age. *Nature*, 546(7657), pp. 293–296.

8. Pedro, N., Brucato, N., Fernandes, V., André, M., Saag, L., Pomat, W., Besse, C., Boland, A., Deleuze, J. F., Clarkson, C., and Sudoyo, H., 2020. Papuan mitochondrial genomes and the settlement of Sahul. *Journal of Human Genetics*, 65:875–887.

9. Jacobs, G. S., Hudjashov, G., Saag, L., Kusuma, P., Darusallam, C. C., Lawson, D. J., Mondal, M., Pagani, L., Ricaut, F. X., Stoneking, M., and Metspalu, M., 2019. Multiple deeply divergent Denisovan ancestries in Papuans. *Cell*, 177(4), pp. 1010–1021.

10. Reich, D., 2018. *Who We Are and How We Got Here: Ancient DNA and the New Science of the Human Past*. Oxford University Press.

11. Keller, A., Graefen, A., Ball, M., Matzas, M., Boisguerin, V., Maixner, F., Leidinger, P., Backes, C., Khairat, R., Forster, M., and Stade, B., 2012. New insights into the Tyrolean Iceman's origin and phenotype as inferred by whole-genome sequencing. *Nature Communications*, 3(1), pp. 1–9.

12. Murphy Jr, W. A., Nedden, D. Z., Gostner, P., Knapp, R., Recheis, W., and Seidler, H., 2003. The iceman: Discovery and imaging. *Radiology*, 226(3), pp. 614–629.

13. Huerta-Sánchez, E., Jin, X., Bianba, Z., Peter, B. M., Vinckenbosch, N., Liang, Y., Yi, X., He, M., Somel, M., Ni, P., and Wang, B., 2014. Altitude adaptation in Tibetans caused by introgression of Denisovan-like DNA. *Nature*, 512(7513), pp. 194–197.

14. Rollo, F., Ubaldi, M., Ermini, L., and Marota, I., 2002. Ötzi's last meals: DNA analysis of the intestinal content of the Neolithic glacier mummy from the Alps. *Proceedings of the National Academy of Sciences*, 99(20), pp. 12594–12599.

15. Perry, G. H., Kistler, L., Kelaita, M. A., and Sams, A. J., 2015. Insights into hominin phenotypic and dietary evolution from ancient DNA sequence data. *Journal of Human Evolution*, 79, pp. 55–63.

16. Mascher, M., Schuenemann, V. J., Davidovich, U., Marom, N., Himmelbach, A., Hübner, S., Korol, A., David, M., Reiter, E., Riehl, S., and Schreiber, M., 2016. Genomic analysis of 6,000-year-old cultivated grain illuminates the domestication history of barley. *Nature Genetics*, 48(9), pp. 1089–1093; Bilgic, H., Hakki, E. E., Pandey, A., Khan, M. K., and Akkaya, M. S., 2016. Ancient DNA from 8400-year-old Çatalhöyük wheat: Implications for the origin of Neolithic agriculture. *PLoS One*,

11(3), p. e0151974; Fornaciari, R., Fornaciari, S., Francia, E., Mercuri, A. M., and Arru, L., 2018. Panicum spikelets from the Early Holocene Takarkori rockshelter (SW Libya): Archaeo-molecular and-botanical investigations. *Plant Biosystems—An International Journal Dealing with All Aspects of Plant Biology*, *152*(1), pp. 1–13.

17. Foley, B. P., Hansson, M. C., Kourkoumelis, D. P., and Theodoulou, T. A., 2012. Aspects of ancient Greek trade re-evaluated with amphora DNA evidence. *Journal of Archaeological Science*, *39*(2), pp. 389–398.

18. Ramos-Madrigal, J., Smith, B. D., Moreno-Mayar, J. V., Gopalakrishnan, S., Ross-Ibarra, J., Gilbert, M.T.P., and Wales, N., 2016. Genome sequence of a 5,310-year-old maize cob provides insights into the early stages of maize domestication. *Current Biology*, *26*(23), pp. 3195–3201.

19. Kistler, L., Maezumi, S. Y., Gregorio de Souza, J., Przelomska, N.A.S., Costa, F. M., Smith, O., Loiselle, H., Ramos-Madrigal, J., Wales, N., Ribeiro, E. R., Morrison, R. R., Grimaldo, C., Prous, A. P., Arriaza, B., Gilbert, M.T.P., Freitas, F.D.O., and Allaby, R. G., 2020. Multiproxy evidence highlights a complex evolutionary legacy of maize in South America. *Science*, *362* (6420), pp. 1309–1313.

20. Hollemeyer, K., Altmeyer, W., Heinzle, E., and Pitra, C., 2008. Species identification of Oetzi's clothing with matrix-assisted laser desorption/ionization time-of-flight mass spectrometry based on peptide pattern similarities of hair digests. *Rapid Communications in Mass Spectrometry*, *22*(18), pp. 2751–2767.

21. Hawass, Z., Gad, Y. Z., Ismail, S., Khairat, R., Fathalla, D., Hasan, N., Ahmed, A., Elleithy, H., Ball, M., Gaballah, F., and Wasef, S., 2010. Ancestry and pathology in King Tutankhamun's family. *JAMA*, *303*(7), pp. 638–647. But see Lorenzen, E. D., and Willerslev, E., 2010. King Tutankhamun's family and demise. *JAMA*, *303*(24), pp. 2471–2475.

22. Stone, A. C., and Ozga, A. T., 2019. Ancient DNA in the study of ancient disease. In *Ortner's Identification of Pathological Conditions in Human Skeletal Remains*. Academic Press, pp. 183–210.

23. Corthals, A., Koller, A., Martin, D. W., Rieger, R., Chen, E. I., Bernaski, M., Recagno, G., and Dávalos, L. M., 2012. Detecting the immune system response of a 500-year-old Inca mummy. *PloS One*, *7*(7), p. e41244.

24. Hansson, G. K., and Edfeldt, K., 2005. Toll to be paid at the gateway to the vessel wall. *Arteriosclerosis Thrombosis and Vascular Biology*, *25*, pp. 1085–1087.

25. Ryg-Cornejo, V., Ioannidis, L. J., Ly, A., Chiu, C. Y., Tellier, J., Hill, D. L., Preston, S. P., Pellegrini, M., Yu, D., Nutt, S. L., Kallies, A., and Hansen, D. S., 2016. Severe malaria infections impair germinal center responses by inhibiting T follicular helper cell differentiation. *Cell Reports*, *14*(1), pp. 68–81.

26. Zeberg, H., and Pääbo, S., 2020. The major genetic risk factor for severe COVID-19 is inherited from Neanderthals. *Nature*, *587*, pp. 610–612.

27. Newbury, D. F., et al. 2009. CMIP and ATP2C2 modulate phonological short-term memory in language impairment. *American Journal of Human Genetics*, 85(2), pp. 264–272.

28. Morris, D., 1967. *The Naked Ape: A Zoologist's Study of the Human Animal.* Random House.

29. Lalueza-Fox, C., Römpler, H., Caramelli, D., Stäubert, C., Catalano, G., Hughes, D., Rohland, N., Pilli, E., Longo, L., Condemi, S., and de La Rasilla, M., 2007. A melanocortin 1 receptor allele suggests varying pigmentation among Neanderthals. *Science*, 318(5855), pp. 1453–1455.

30. Crawford, N. G., et al. 2017. Loci associated with skin pigmentation identified in African populations. *Science*, 358(6365).

31. Pinhasi, R., Fernandes, D., Sirak, K., Novak, M., Connell, S., Alpaslan-Roodenberg, S., Gerritsen, F., Moiseyev, V., Gromov, A., Raczky, P., and Anders, A., 2015. Optimal ancient DNA yields from the inner ear part of the human petrous bone. *PLoS One*, 10(6), p. e0129102.

32. Sirak, K., Fernandes, D., Cheronet, O., Harney, E., Mah, M., Mallick, S., Rohland, N., Adamski, N., Broomandkhoshbacht, N., Callan, K., Candilio, F., Lawson, A. M., Mandl, K., Oppenheimer, J., Stewardson, K., Zalzala,F., Anders, A., Bartík, J., Coppa, A., Dashtseveg, T., Évinger, S., Farkaš, Z., Hajdu, T., Bayarsaikhan,J., McIntyre, L., Moiseyev, V., Okumura, M., Pap, I., Pietrusewsky, M., Raczky, P., Šefčáková, A., Soficaru, A., Szeniczey,T., Szőke, B. M., Gerven, D. V., Vasilyev, S., Bell, L., Reich, D., and Pinhasi, R., 2020. Human auditory ossicles as an alternative optimal source of ancient DNA. *Genome Research*, 30(3), pp. 427–436.

33. Hansen, H. B., Damgaard, P. B., Margaryan, A., Stenderup, J., Lynnerup, N., Willerslev, E., and Allentoft, M. E., 2017. Comparing ancient DNA preservation in petrous bone and tooth cementum. *PLoS One*, 12(1), p. e0170940.

34. Wade, L., 2015. Breaking a tropical taboo. *Science*, 349(6246), pp. 370–371. https://doi.org/10.1126/science.349.6246.370.

35. Sutikna, T., Tocheri, M. W., Faith, J. T., Awe, R. D., Meijer, H. J., Saptomo, E. W., and Roberts, R. G., 2018. The spatio-temporal distribution of archaeological and faunal finds at Liang Bua (Flores, Indonesia) in light of the revised chronology for Homo floresiensis. *Journal of Human Evolution*, 124, pp. 52–74.

36. Reich, D., 2018. *Who We Are and How We Got Here: Ancient DNA and the New Science of the Human Past.* Oxford University Press.

37. Brumm, A., Van Den Bergh, G. D., Storey, M., Kurniawan, I., Alloway, B. V., Setiawan, R., Setiyabudi, E., Grün, R., Moore, M. W., Yurnaldi, D., Puspaningrum, M. R., Wibowo, U. P., Insani, H., Sutisna, I., Westgate, J. A., Pearce, N.J.G., Duval, M., Meijer, H.J.M., Aziz, F., Sutikna, T., van der Kaars, S., Flude S., and Morwood, M. J., 2016. Age and context of the oldest known hominin fossils from Flores. *Nature*, 534(7606), pp. 249–253.

Chapter 11. Plants

1. Miao, L., Moczydłowska, M., Zhu, S., and Zhu, M., 2019. New record of organic-walled, morphologically distinct microfossils from the late Paleoproterozoic Changcheng Group in the Yanshan Range, North China. *Precambrian Research, 321*, pp. 172–198.

2. Javaux, E. J., Marshall, C. P., and Bekker, A., 2010. Organic-walled microfossils in 3.2-billion-year-old shallow-marine siliciclastic deposits. *Nature, 463*(7283), pp. 934–938.

3. Glauser, A. L., Harper, C. J., Taylor, T. N., Taylor, E. L., Marshall, C. P., and Marshall, A. O., 2014. Reexamination of cell contents in Pennsylvanian spores and pollen grains using Raman spectroscopy. *Review of Palaeobotany and Palynology, 210*, pp. 62–68; Wall, D., 1965. Microplankton, pollen, and spores from the Lower Jurassic in Britain. *Micropaleontology, 11*(2), pp. 151–190.

4. Fastovsky, D. E., and Bercovici, A., 2016. The Hell Creek Formation and its contribution to the Cretaceous–Paleogene extinction: A short primer. *Cretaceous Research, 57*, pp. 368–390.

5. Mackenzie, G., Boa, A. N., Diego-Taboada, A., Atkin, S. L., and Sathyapalan, T., 2015. Sporopollenin, the least known yet toughest natural biopolymer. *Frontiers in Materials, 2*, p. 66.

6. Li, F. S., Phyo, P., Jacobowitz, J., Hong, M., and Weng, J. K., 2019. The molecular structure of plant sporopollenin. *Nature Plants, 5*(1), pp. 41–46.

7. Fraser, W. T., Scott, A. C., Forbes, A.E.S., Glasspool, I. J., Plotnick, R. E., Kenig, F., and Lomax, B. H., 2012. Evolutionary stasis of sporopollenin biochemistry revealed by unaltered Pennsylvanian spores. *New Phytologist, 196*(2), pp. 397–401.

8. Taylor, E. L., Taylor, T. N., and Krings, M., 2009. *Paleobotany: The Biology and Evolution of Fossil Plants*. Academic Press; D'Angelo, J. A., Lyons, P. C., Mastalerz, M., and Zodrow, E. L., 2013. Fossil cutin of Macroneuropteris scheuchzeri (Late Pennsylvanian seed fern, Canada). *International Journal of Coal Geology, 105*, pp. 137–140.

9. Soul, L. C., Barclay, R. S., Bolton, A., and Wing, S. L., 2019. Fossil Atmospheres: A case study of citizen science in question-driven palaeontological research. *Philosophical Transactions of the Royal Society B, 374*(1763), p. 20170388.

10. Tralau, H., 1968. Evolutionary trends in the genus Ginkgo. *Lethaia, 1*(1), pp. 63–101.

11. McInerney, F. A., and Wing, S. L., 2011. The Paleocene-Eocene thermal maximum: A perturbation of carbon cycle, climate, and biosphere with implications for the future. *Annual Review of Earth and Planetary Sciences, 39*, pp. 489–516.

12. Berbee, M. L., Strullu-Derrien, C., Delaux, P. M., Strother, P. K., Kenrick, P., Selosse, M. A., and Taylor, J. W., 2020. Genomic and fossil windows into the secret lives of the most ancient fungi. *Nature Reviews Microbiology, 18*(12), pp. 717–730.

13. Grattan, D. W., 1991. The conservation of specimens from the Geodetic Hills fossil forest site, Canadian Arctic Archipelago. In *Tertiary Fossil Forests of the Geodetic Hills, Axel Heiberg Island, Arctic Archipelago*. Geological Survey of Canada, Bulletin 403, pp. 213–227.

14. Marynowski, L., Bucha, M., Smolarek, J., Wendorff, M., and Simoneit, B. R., 2018. Occurrence and significance of mono-, di-and anhydrosaccharide biomolecules in Mesozoic and Cenozoic lignites and fossil wood. *Organic Geochemistry, 116*, pp. 13–22.

15. Lücke, A., Helle, G., Schleser, G. H., Figueiral, I., Mosbrugger, V., Jones, T. P., and Rowe, N. P., 1999. Environmental history of the German Lower Rhine Embayment during the Middle Miocene as reflected by carbon isotopes in brown coal. *Palaeogeography, Palaeoclimatology, Palaeoecology, 154*(4), pp. 339–352; Jahren, A. H., and Sternberg, L.S.L., 2002. Eocene meridional weather patterns reflected in the oxygen isotopes of Arctic fossil wood. *GSA Today, 12*(1), pp. 4–9.

16. Marynowski, L., Bucha, M., Smolarek, J., Wendorff, M., and Simoneit, B. R., 2018. Occurrence and significance of mono-, di-and anhydrosaccharide biomolecules in Mesozoic and Cenozoic lignites and fossil wood. *Organic Geochemistry, 116*, pp. 13–22.

17. Griffith, J. D., Willcox, S., Powers, D. W., Nelson, R., and Baxter, B. K., 2008. Discovery of abundant cellulose microfibers encased in 250 Ma Permian halite: A macromolecular target in the search for life on other planets. *Astrobiology, 8*(2), pp. 215–228.

18. Vreeland, R. H., Rosenzweig, W. D., and Powers, D. W., 2000. Isolation of a 25-million-year-old halotolerant bacterium from a primary salt crystal. *Nature, 407*(6806), pp. 897–900.

19. Nickle, D. C., Learn, G. H., Rain, M. W., Mullins, J. I., and Mittler, J. E., 2002. Curiously modern DNA for a "250-million-year-old" bacterium. *Journal of Molecular Evolution, 54*(1), pp. 134–137; Hebsgaard, M. B., Phillips, M. J., and Willerslev, E., 2005. Geologically ancient DNA: Fact or artefact? *Trends in Microbiology, 13*(5), pp. 212–220.

20. Stein, W. E., Mannolini, F., Hernick, L. V., Landing, E., and Berry, C. M., 2007. Giant cladoxylopsid trees resolve the enigma of the Earth's earliest forest stumps at Gilboa. *Nature, 446*(7138), pp. 904–907.

21. Yang, H., Huang, Y., Leng, Q., LePage, B. A., and Williams, C. J., 2005. Biomolecular preservation of Tertiary Metasequoia fossil lagerstätten revealed by comparative pyrolysis analysis. *Review of Palaeobotany and Palynology, 134*(3–4), pp. 237–256.

22. Yang, H., Huang, Y., Leng, Q., LePage, B. A., and Williams, C. J., 2005. Biomolecular preservation of Tertiary Metasequoia fossil lagerstätten revealed by comparative pyrolysis analysis. *Review of Palaeobotany and Palynology, 134*(3–4),

pp. 237–256; Logan, G. A., Boon, J. J., and Eglinton, G., 1993. Structural biopolymer preservation in Miocene leaf fossils from the Clarkia site, northern Idaho. *Proceedings of the National Academy of Sciences, 90*(6), pp. 2246–2250.

23. Dong, W., Tan, L., Zhao, J., Hu, R., and Lu, M., 2015. Characterization of fatty acid, amino acid and volatile compound compositions and bioactive components of seven coffee (*Coffea robusta*) cultivars grown in Hainan Province, China. *Molecules, 20*(9), pp. 16687–16708.

24. Uehara, A., Nakata, M., Kitajima, J., and Iwashina, T., 2012. Internal and external flavonoids from the leaves of Japanese Chrysanthemum species (Asteraceae). *Biochemical Systematics and Ecology, 41*, pp. 142–149.

25. Lange, B. M., 2015. The evolution of plant secretory structures and emergence of terpenoid chemical diversity. *Annual Review of Plant Biology, 66*, pp. 139–159.

26. Bray, P. S., and Anderson, K. B., 2009. Identification of Carboniferous (320 million years old) class Ic amber. *Science, 326*(5949), pp. 132–134.

27. Simoneit, B. R., Otto, A., Kusumoto, N., and Basinger, J. F., 2016. Biomarker compositions of *Glyptostrobus* and *Metasequoia* (Cupressaceae) fossils from the Eocene Buchanan Lake Formation, Axel Heiberg Island, Nunavut, Canada reflect diagenesis from terpenoids of their related extant species. *Review of Palaeobotany and Palynology, 235*, pp. 81–93; Otto, A., Simoneit, B. R., and Rember, W. C., 2003. Resin compounds from the seed cones of three fossil conifer species from the Miocene Clarkia flora, Emerald Creek, Idaho, USA, and from related extant species. *Review of Palaeobotany and Palynology, 126*(3–4), pp. 225–241.

28. Giannasi, D. E., and Niklas, K. J., 1981. Comparative paleobiochemistry of some fossil and extant Fagaceae. *American Journal of Botany, 68*(6), pp. 762–770.

29. Zhao, Y. X., Li, C. S., Luo, X. D., Wang, Y. F., and Zhou, J., 2006. Palaeophytochemical constituents of Cretaceous *Ginkgo coriacea* Florin leaves. *Journal of Integrative Plant Biology, 48*(8), pp. 983–990.

30. Crown, P. L., and Hurst, W. J., 2009. Evidence of cacao use in the Prehispanic American Southwest. *Proceedings of the National Academy of Sciences, 106*(7), pp. 2110–2113.

Chapter 12. The Future of Studying the Past

1. Stern, J. C., Sutter, B., Freissinet, C., Navarro-González, R., McKay, C. P., Archer, P. D., Buch, A., Brunner, A. E., Coll, P., Eigenbrode, J. L., and Fairen, A. G., 2015. Evidence for indigenous nitrogen in sedimentary and aeolian deposits from the Curiosity rover investigations at Gale crater, Mars. *Proceedings of the National Academy of Sciences, 112*(14), pp. 4245–4250.

2. Greaves, J. S., Richards, A. M., Bains, W., Rimmer, P. B., Sagawa, H., Clements, D. L., Seager, S., Petkowski, J. J., Sousa-Silva, C., Ranjan, S., and Drabek-Maunder,

E., 2020. Phosphine gas in the cloud decks of Venus. *Nature Astronomy*, 5(7), pp. 655–664.

3. Schopf, J. W., Kitajima, K., Spicuzza, M. J., Kudryavtsev, A. B., and Valley, J. W., 2018. SIMS analyses of the oldest known assemblage of microfossils document their taxon-correlated carbon isotope compositions. *Proceedings of the National Academy of Sciences*, 115(1), pp. 53–58.

4. Liberles, D. A., ed., 2007. *Ancestral Sequence Reconstruction*. Oxford University Press.

5. Wilson, C., et al. 2015. Kinase dynamics. Using ancient protein kinases to unravel a modern cancer drug's mechanism. *Science*, 347:882–886.

6. Ugalde, J. A., Chang, B. S., and Matz, M. V., 2004. Evolution of coral pigments recreated. *Science*, 305(5689), pp. 1433–1433.

7. Akanuma, S., Nakajima, Y., Yokobori, S. I., Kimura, M., Nemoto, N., Mase, T., Miyazono, K. I., Tanokura, M., and Yamagishi, A., 2013. Experimental evidence for the thermophilicity of ancestral life. *Proceedings of the National Academy of Sciences*, 110(27), pp. 11067–11072.

8. Gaucher, E. A., Thomson, J. M., Burgan, M. F., and Benner, S. A., 2003. Inferring the palaeoenvironment of ancient bacteria on the basis of resurrected proteins. *Nature*, 425(6955), pp. 285–288.

9. Ayuso Fernández, I., 2019. Resurrection of ancestral ligninolytic peroxidases (doctoral dissertation, Universidad Complutense de Madrid).

10. Nguyen, A., Hoang, D. M., Nguyen, T.A.M., Nguyen, D. T., Long, B., Meijaard, E., Holland, J., Wilting, A., and Tilker, A., 2019. Camera-trap evidence that the silver-backed chevrotain Tragulus versicolor remains in the wild in Vietnam. *Nature Ecology & Evolution*, 3(12), pp. 1650–1654.

11. Hu, H. H., and Cheng, W., 1948. On the new family Metasequoiaceae and on *Metasequoia glyptostroboides*: A living species of the genus *Metasequoia* found in Szechuan and Hupeh. *Bull. Fan Memorial Inst. Biol., New Series 1*: 153–161.

12. Yashina, S., Gubin, S., Maksimovich, S., Yashina, A., Gakhova, E., and Gilichinsky, D., 2012. Regeneration of whole fertile plants from 30,000-y-old fruit tissue buried in Siberian permafrost. *Proceedings of the National Academy of Sciences*, 109(10), pp. 4008–4013.

13. Yamagata, K.,. et al., 2019. Signs of biological activities of 28,000-year-old mammoth nuclei in mouse oocytes visualized by live-cell imaging. *Scientific Reports*, 9(1), pp. 1–12.

14. Colson, P., De Lamballerie, X., Yutin, N., Asgari, S., Bigot, Y., Bideshi, D. K., Cheng, X. W., Federici, B. A., Van Etten, J. L., Koonin, E. V., and La Scola, B., 2013. "Megavirales," a proposed new order for eukaryotic nucleocytoplasmic large DNA viruses. *Archives of Virology*, 158(12), pp. 2517–2521.

15. Legendre, M., Bartoli, J., Shmakova, L., Jeudy, S., Labadie, K., Adrait, A., Lescot, M., Poirot, O., Bertaux, L., Bruley, C., and Couté, Y., 2014. Thirty-thousand-year-old distant relative of giant icosahedral DNA viruses with a pandoravirus morphology. *Proceedings of the National Academy of Sciences*, 111(11), pp. 4274–4279.

16. Hutchison, C. A., Chuang, R. Y., Noskov, V. N., Assad-Garcia, N., Deerinck, T. J., Ellisman, M. H., Gill, J., Kannan, K., Karas, B. J., Ma, L., Pelletier, J. F., Qi, Z.-Q., Richter, R. A., Strychalski, E. A., Sun, L., Suzuki, Y., Tsvetanova, B., Wise, K. S., Smith, H. O., Glass, J. L., Merryman, C., Gibson, D. G., and Venter, J. C. 2016. Design and synthesis of a minimal bacterial genome. *Science*, *351*(6280).

17. Wray, B., and Church, G., 2019. *Rise of the Necrofauna: The Science, Ethics, and Risks of De-Extinction*. Greystone Books.

ILLUSTRATION CREDITS

1.1. Photograph by Dale Greenwalt.

1.2. Photograph by Dale Greenwalt; USNM PAL 559050; Blood-engorged mosquito. Courtesy of the Smithsonian Institution. © 2022 National Academy of Sciences.

2.1. Photograph by Dale Greenwalt.

2.2. Photograph by Dale Greenwalt; USNM PAL 311313; Horodyskia. Courtesy of the Smithsonian Institution.

2.3. Photograph by Dale Greenwalt; USNM PAL 773712; Mazon creek nodule. Courtesy of the Smithsonian Institution.

2.4. Photograph by Alan Munro.

3.1. Photograph by Dale Greenwalt; USNM PAL S5996 A-D; Rhodocrinites kirbyi (crinoid). Courtesy of the Smithsonian Institution.

3.2. Photograph by Dale Greenwalt.

3.3. Photograph by Dale Greenwalt.

3.4. © Denver Museum of Nature & Science.

3.5. Photograph by Dale Greenwalt.

3.6. Photograph by Dale Greenwalt.

4.1. Right photograph courtesy of Professor Ren, College of Life Sciences, Capital Normal University, Beijing, China; left photograph by Dale Greenwalt, courtesy of the Entomology Department, National Museum of Natural History, Smithsonian Institution.

4.2. Drawing by Dale Greenwalt.

5.1. Drawing by Dale Greenwalt.

6.1. Photograph by Dale Greenwalt; USNM PAL 595153; Staphylinid beetle. Courtesy of the Smithsonian Institution.

6.2. Drawing by Dale Greenwalt.

7.1. Drawing by Dale Greenwalt.

8.1. Courtesy of Surmik, D., et al., 2016. Spectroscopic studies on organic matter from Triassic reptile bones, Upper Silesia, Poland. *PLoS One, 11*(3): e0151143. https://doi.org/10.1371/journal.pone.0151143. The figures were altered to remove amorphous background material and to group three vessels in a single figure.

9.1. Courtesy of Nel, A., and Coty, D., 2016. A fossil dung midge in Mexican amber (Diptera: Scatopsidae). *Palaeontologia Electronica, 19*, pp. 1–6.

9.2. Photograph by Dale Greenwalt.

9.3. Drawing by Dale Greenwalt.

9.4. Courtesy of the NASA/Goddard Space Flight Center Scientific Visualization Studio.

9.5. Courtesy of Vicente Rodríguez, A., Villalba Breva, S., Ferràndez i Cañadell, C., and Martín-Closas, C., 2016. Revision of the Maastrichtian-Palaeocene charophyte biostratigraphy of the Fontllonga reference section (southern Pyrenees, Catalonia, Spain). *Geologica Acta, 14,* pp. 349–362. CC BY-SA.

11.1. National Museum of Natural History specimen # USNM PAL 772399; photograph by Rich Barclay.

11.2. Specimen courtesy of Department of Geological Sciences, University of Saskatchewan, Saskatoon, Canada; photograph by Dale Greenwalt.

11.3. Courtesy of Dong, W., Tan, L., Zhao, J., Hu, R., and Lu, M., 2015. Characterization of fatty acid, amino acid and volatile compound compositions and bioactive components of seven coffee (*Coffea robusta*) cultivars grown in Hainan Province, China. *Molecules, 20*(9), pp. 16687–16708. This figure has been redrawn from the original figures 3B-D. CC BY 4.0.

12.1. Drawing by Dale Greenwalt.

INDEX

A page number in italics refers to a figure.